Heublein   Transmitting Data
without Interference

Transmitting Data without Interference

Cables in Building Installations and in Industrial Measurement and Process Control

Hans Heublein

Publicis MCD Verlag

Die Deutsche Bibliothek – CIP-Einheitsaufnahme

Heublein, Hans:
Transmitting data without interference : cables in building
installations and in industrial measurement and process control / Hans
Heublein. [Issued by Siemens-Aktiengesellschaft, Berlin and
Munich]. – Erlangen ; München : Publicis-MCD-Verl., 1998
    ISBN 3-89578-073-1

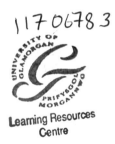

This book was carefully produced. Nevertheless, authors, editors and publisher do not warrant
the information contained therein to be free of errors. Readers are advised to keep in mind that
statements, data, illustrations, procedural details or other items may inadvertently be inaccurate.
Terms reproduced in this book may be registered trademarks, the use of which by third parties
for their own purposes may violate the rights of the owners of those trademarks.

ISBN 3-89578-073-1

Issued by Siemens Aktiengesellschaft, Berlin and Munich
Published by Publicis MCD Verlag, Erlangen and Munich
© 1998 by Publicis MCD Werbeagentur GmbH, Munich

Introduction

In the past the electrical wiring within building installations was solely influenced by power supply technology. Digital electronics and information technology have meanwhile moved to the forefront. Examples being: building automation and control systems, air-conditioning systems, security and fire protection systems, and especially telecommunication and data processing systems.

Even though cable standards continue to strictly differentiate between power supply cables and telecommunication or data processing cables, their application areas increasingly overlap. Upon publication of the standards for generic cabling systems (e.g. ISO/IEC DIS 11801) and with market acceptance of the European Installation Bus (EIB), electricians will, at the latest, begin to install networks for computer and control systems along with conventional power distribution networks. The planning and installation of cable networks requires technical knowledge of signal transmission as well as power supply technology. After all, data and communication cables are often necessary within power plants or power supply applications. Furthermore, the legal aspects of the electromagnetic compatibility (EMC) should also be considered during planning and installation phases.

Besides cables, the electrical characteristics of the interconnected components or systems determine the transmission response. Only after all elements of the entire transmission system are coordinated with each other, can an optimal transmission response ensue. During planning and commissioning of transmission paths or networks, this demands a broad understanding of various electrical disciplines, such as: electronic circuit theory, high frequency technology, electromagnetics, cable technology, power engineering, etc.

In contrast to building systems technology, automation technology (industrial control and instrumentation systems) was confronted early with a comprehensive, stormy switch to electronics. Additional mechanical robustness is hereby demanded from the cables and their materials. These features are largely determined by the chemical, thermal and kinetic conditions of the application environment.

This book provides an overview of the various technical disciplines and their combined effects. Therefore, the individual technical specialties are not discussed in their entirety.

Erlangen, November 1997   Publicis MCD Verlag

Table of Contents

1 Fundamentals of Signal Transmission

Semiconductor devices, such as diodes and transistors, are mainly used in power supply technology as fast and reliable switching elements and are considered user-friendly system components. Preferably, these components are only comprised of elements having identical circuit technology.

Integrated circuit technologies have been progressively developed. Their technological differences stem from the demands of system designers and the available manufacturing capabilities during the developmental stages. In general, contact technology was succeeded by analog technology, to be superseded by digital technology. Similarly, the transmission frequencies became successively higher. Simultaneously, the applied signal voltages changed from relatively high AC (alternating current) to low DC (direct current) voltages.

1.1 Analog Technology

Analog technology has largely been overtaken by digital technology. Nonetheless, in future it will remain a suitable alternative for older systems and special applications. The best known analog systems are:

± 10 V systems,
± 20 mA systems and the
4...20 mA systems.

The analog signal for the above mentioned systems is usually transmitted unidirectionally via two-conductor copper cables. Information is transmitted with either a variable DC voltage or variable DC current. The signal's amplitude corresponds to the value of the transmitted information. Therefore, distortions or perturbations influencing the signal's integrity cannot be corrected afterwards.

1.1.1 The Operational Amplifier as a Significant Component in Analog Technology

The primary component for all analog systems is the operational amplifier (OP). Diagram 1.1 depicts its circuit symbol and voltage designations.

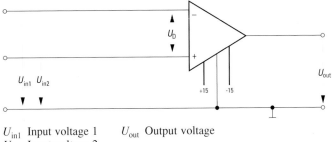

U_{in1} U_{in2}

U_{in1} Input voltage 1 U_{out} Output voltage
U_{in2} Input voltage 2

Diagram 1.1   Circuit symbol and voltage designation for an operational amplifier

Operational amplifiers usually have a supply voltage of either ± 15 V DC or ± 5 V DC. The reference potential for all voltages is the neutral conductor N of the supply voltage.

The output voltage U_{out} of an unwired operational amplifier is proportional to the difference between the two input voltage levels (differential voltage U_D). Depending on the tolerance, deviations may exist. The zero input voltage (offset voltage) is understood as the voltage difference to be applied between both inputs of the unwired amplifier in order to achieve a 0 V output. The operational amplifier is wired with other resistors and components to acquire its desired performance characteristics. The operating signal is then applied as a differential voltage U_D to both inputs of the operational amplifier. In the absence of interference, the output voltage of the amplifier is an undistorted representation of the input signal.

The following values are typical for an operational amplifier:

(Example: OPA1620 from BURR BROWN)

Supply voltage V_{cc}	typ.	±5 V
	min.	±4 V
	max.	±6 V
Output voltage (load resistance: 100 Ω, V_{cc}: ±5 V)	typ.	±3 V
Input common-mode voltage range	typ.	±3.5 V
Voltage amplification (load resistance: 100 Ω)	typ.	60 dB
Common-mode rejection (*CMR*), (V_{cc}: ±5 V)	typ.	75 dB
Offset voltage, input	typ.	±200 µV

The maximum switching speeds of an op amp are very high (several MHz). However, typical systems usually require much lower speeds. Since slower processing speeds are less sensitive to interference, the operating speeds of the amplifier are often reduced, with the addition of discrete components, as far as the frequency of the input signal allows.

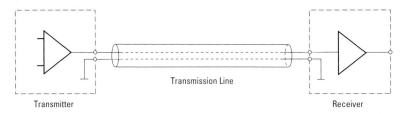

Receiver

Diagram 1.2 Circuit diagram of an analog signal transmission

In signal transmission applications, the op amp is employed as an output amplifier in the transmitter and as an input amplifier in the receiver. By wiring them with different components, both amplifiers are given specific characteristics. The input and output amplifiers are linked via a transmission cable (diagram 1.2). An output amplifier having this function is known as a line driver.

1.1.2 Processing Interference Signals in Analog Technology

Diagram 1.3 depicts the differential voltage U_D and the output voltage U_{out} of an op amp wired within a circuit.

The common-mode voltage U_C is derived from the arithmetic midpoint between the two input voltages U_{in1} and U_{in2} (diagram 1.1):

$$U_C = \frac{U_{in1} + U_{in2}}{2} \tag{1}$$

An op amp can experience distortion at either input in the following manner:

▷ *Interference voltage at one input*

 Hereby, the voltage difference between both inputs is changed and consequently also the output voltage.

▷ *Dissimilar interference voltages at both inputs*

 Similarly, the voltage difference between the inputs is changed and thereby the output voltage.

▷ *Identical interference voltages at both inputs*

 Hereby, the voltage difference at the input does not change. Theoretically, undesirable changes in the output voltage should not result. Realistically, the output voltage undergoes a quantifiable alteration. The magnitude of change is a known quantity dependent upon the op amp model chosen and is given in the manufacturer's data sheets. This parameter is called the common-mode rejection (*CMR*). The term "in-phase suppression" is also commonly used. The common-mode rejec-

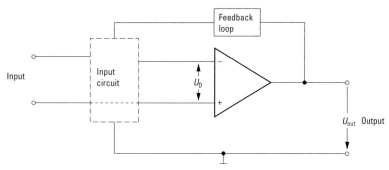

U_D Differential voltage

Diagram 1.3 Voltage designation for a wired operational amplifier

tion is derived from the amplification factors V_{UC} and V_{UD} for both the common-mode voltage U_C and the differential voltage U_D, whereby the output voltage remains identical (U_{out}):

$$CMR = 20 \log (V_{UC} / V_{UD}) \qquad \text{in dB} \qquad (2)$$

The value of the common-mode rejection *CMR* also determines how many times less the common-mode interference is amplified with respect to the operating signal (differential voltage).

Example:

With a common-mode rejection of 80 dB, a common-mode signal of 1 V generates the identical output voltage as a normal-mode signal of 100 μV.

▷ *An overly large common-mode and/or normal-mode voltage*

These conditions may over-drive or disrupt the switching of the op amp (e. g. lightening stroke).

▷ *Voltage loss along the transmission cable*

Due to the voltage drop along the transmission cable, the input signal becomes distorted and may be incorrectly processed by the op amp.

1.1.3 Preventative Measures against Interference in Analog Systems

Principally, the type, installation, wiring, and termination of the transmission cable should not induce any additional spurious common-mode or normal-mode voltages. Distortion that noticeably differs from the operating signal with regard to its frequency, may be simply and easily suppressed at the input of the op amp with a filter. Even though a filter protects the op amp, it is necessary to consider that the cables (conductors and shield) may introduce interference at other locations within the equipment.

When the voltage drop along the cable is too large, the magnitude (amplitude) of the transmitted information is undesirably and severely distorted. Cables with a larger conductor cross-section can counteract this effect. However, due to their greater conductor separation, thicker cables are more sensitive to inductive interference which may result, in turn, potentially distorting normal-mode voltages. Simultaneously, the dissymmetries of the cable's construction within its tolerance boundaries increase and may also result in interfering common-mode voltages. For this reason alone, the transmission of information expressed in voltage amplitudes is restricted to "short" cable lengths. Greater distances can be more effectively and reliably bridged by employing the current as the information carrier. Unfortunately, the above referenced interference effects also limit the cable's length by this method.

Interference at the amplifier's output is barely noticeable at the input because the amplifier is usually wired to sufficiently avoid feedback to the input. However, interference carried along the cable may disturb the neighboring components of the transmitter.

Analog technology increasingly demands higher resolution. Therefore, very small signal voltages in the mV range must be reliably and exactly transmitted without interference. This necessitates costly protection measures within the cable's construction. Analog measurement technology permits a maximum error tolerance of less than 1 %. This translates to a minimum signal-to-noise ratio of 40 dB. In comparison, approx. 10 dB are sufficient for telephony, and values between 30 and 60 dB are typical for radio and television. The well known, typical methods of reconditioning a distorted digital signal or of checking the plausibility of the incoming digital information are not feasible for analog technology. A distorted signal cannot be identified as such and is, therefore, incorrectly further processed.

These facts make evident that for analog technology, noise immunity requirements are largely determined by the signal integrity demanded by the application. The application areas of analog transmission systems are generally determined by the resolution limits and electromagnetic susceptibility of the system for small signals, its non-linearities over the entire range of values, and its temperature dependency. In comparison to larger digital systems, analog systems are significantly more expensive.

1.2 Digital Technology

Digital technology employs highly integrated semiconductor components, discrete components and DC voltages as the information carrier.

1.2.1 Families of Digital Circuits

The electronic systems designer deals with complete modules having predetermined logical functions specified by the manufacturer as a result of the IC technology (integrated circuit technology) employed. The following terms are well-known products of the IC technologies:

RTL-circuits	Resistor-Transistor Logic
DTL-circuits	Diode-Transistor Logic
DCTL-circuits	Direct-Coupled Transistor Logic
LSL-circuits	Low-Speed Logic: DTLZ (Diode-Transistor Logic with Zener-Diode)
TTL-circuits	Transistor-Transistor Logic, with the following subgroups:

TTL	Standard
TTL-LP	Low-Power
TTL-HS	High-Speed
TTL-SC	Schottky
TTL-LS	Low-Power Schottky
TTL-ALS	Advanced-Low-Power-Schottky
TTL-AS	Advanced-Schottky
TTL-F	Fairchild AS, Fast-TTL
TTL-HCT	High-Speed CMOS-TTL
TTL-ACT	Advanced-CMOS-TTL

ECL-circuits	Emitter-Coupled Logic; other common terms: CML, E^2CL or ECTL
P-MOS-circuits	P-channel Metal-Oxide Semiconductor Field-Effect Transistor
N-MOS-circuits	N-channel Metal-Oxide Semiconductor Field-Effect Transistor
C-MOS-circuits	Complementary Metal-Oxide Semiconductor Field Effect Transistor; other common terms: COS-MOS
C- MOS-HC	High-Speed-CMOS
C-MOS-AC	Advanced-CMOS

RTL- and DTL-circuits belong to the first generation of electronic circuits. They are no longer employed. Modern circuits predominantly employ either CMOS-, MOS-, ECL- or TTL-technologies.

Within a circuit arrangement, usually circuits belonging to a single family are employed. Only in rare cases or for application-specific solutions, is it essential to resort to other circuit technologies. For example, interference or transmission dependent conditions often necessitate switching to another circuit family at the interface of the module. Such cases are always associated with extra efforts for electrical adaptations. In signal transmission applications, the differences between the individual circuit families are important with regard to the voltage level, maximum switching frequencies and the signal-to-noise ratio.

1.2.2 Parameters of Electronic Circuits

Operating voltage (supply voltage)

The parameters of an electronic component are valid only for a given operating voltage or range. Even momentary deviations in the ns-range may lead to an unspecified behaviour of the circuit. Instead of the nominal value, the operating voltage is often expressed as a minimum, maximum, or typical value.

Input level

The levels H (HIGH) and L (LOW) for binary information are each assigned to a specific voltage *value* or *range* in digital electronics. If the input signal lies outside of this range, the function of the circuit is indeterminate.

Output level

The binary output signal of the circuit must also lie within one of two voltage *ranges* to which the logical condition of the circuit is assigned.

The permitted range of the input level of a circuit is almost always different from the permitted range for the output level.

Switching times

The *signal propagation time* t_P expresses the time delay of a circuit between a change in the input signal and its corresponding change in the output signal. It is also known as the signal delay time. During a logical shift in the output signal, the voltage may swing between 10% and 90% of its entire range during the *signal transition time* t_T.

Static signal-to-noise ratio

The term *signal-to-noise ratio* is also a customary alternative to the term *noise immunity*. The static signal-to-noise ratio characterizes the behavior of a logic element with regard to disturbances having an influence lasting longer than the mean signal delay time t_p. According to its definition, the signal-to-noise ratio is the maximum voltage which may be superimposed the input signal without changing the switching condition of the circuit. The static signal-to-noise ratio is, therefore, the gap:

▷ between the potentially maximum output voltage of a logic element at the L-level and the maximum input voltage permitted by the activated logic element, so that during this period the signal may be further processed and accepted as the L-level.

or

▷ between the potentially minimum output value of a logic element at the H-level and the required input voltage, so that this signal may be further processed and accepted as the H-level.

These values are valid for all permissible operating conditions and portray the worst-case situation.

The value of the static signal-to-noise ratio is usually significantly smaller than the arithmetical difference between both specified voltage ranges.

Typical signal-to-noise ratio

The *typical signal-to-noise ratio* stated in the data sheets is usually given for a limited range within the allowable, specified operating conditions for the op-amp (e.g. smaller temperature ranges). It is not valid for worst-case conditions, especially those not uniformly established.

Dynamic signal-to-noise ratio

The *dynamic signal-to-noise ratio* (also referred to as *dynamic noise immunity*) describes the behaviour of a logic element with regard to spurious interference lasting for a duration much shorter than the mean signal delay time t_P of the logic element. In comparison to the static signal-to-noise ratio, the energy introduced by the spurious signal is definitive rather than its voltage. In order to switch the condition of a logic element, disturbances (or operating signals) of such a small impulse duration must demonstrate a significantly higher amplitude than static disturbances. As the impulse duration increases, the value of the dynamic signal-to-noise ratio begins to approach that of the static signal-to-noise ratio.

Table 1.1 contains interesting values for various circuit families. This table is incomplete, because a portion of the terminology and definitions have either just recently been established or have undergone a certain transformation making them no longer easily comparable. Exact, valid values are listed only in the respective manufacturer's catalog. Therefore, this table is restricted to the magnitudes and differences between the various circuit families.

Table 1.1 Summary of circuit parameters

Circuit family	Operating voltage	Signal level Inputs H/L	Outputs H/L	Mean signal delay time	Maximum switching frequency	Static noise immunity H/L
	typical V	min./max. V	typical V	ns	MHz	typical V
RTL	5			10		0.3
DTL	6	3.6/1.4	5.0 / 0.5	30		1.2
LSL	12 or 15	7.5/4.5	14/1.0	195		8 / 5
TTL	5	2.0/0.8	2.4…2.8/ 0.2…0.4	10	50	1
TTL-LP	5			33	3	1
TTL-HS	5			5	80	1
TTL-LS	5			9.5	50	0.7/0.3
TTL-SC	5			3	130	0.5
TTL-ALS	5			5		0.7/0.3
TTL-F	5			2		0.7/0.3
TTL-AS	5			1.7		0.7/0.3
TTL-HCT	5			10		2.4/0.7
TTL-ACT	5			4		2.4/0.7
CMOS-HC	2 4.5 5.5			6.4		0.4/0.2 1.25/0.8 1.7/1.1
CMOS-AC	3 4.5 5.5			2.5		0.8 1.25 1.55
ECL	−5	−1…0/ −5…1.4	−0.8/1.6	2	250	0.3
P-MOS	−12	−1…0.3/ −12…−6	−1/−11	80	2	3
N-MOS	5	2.2/0.65	3.5/0.3	15	20	2
C-MOS	3 5 10 15	3…5/ 0…2 7…10/ 0…3	4.99…5.0/ 0…0.01 9.99…10/ 0…0.01	25	10	2

L LOW-level   H HIGH-level

Table 1.2 Destructive voltage limits for semiconductor components

Component type	Destructive voltage limits V
V-MOS	50…1800
MOSFET	100… 200
EPROM	100… 500
Junction-FET	150…1600
Analog IC (FET)	150… 500
Analog IC (bipolar)	200…2500
CMOS	250…2000
Schottky-Diode	300…3000
Schottky-IC	300…2500
Transistors (bipolar)	400…7000
Thyristors, Triac	700…2500
Surface Wave Filters	150… 500
Film Resistance	300…3000
ECL	500…1500

(Source: Siemens Matsushita Components)

Table 1.2 lists the destructive voltage limits for various semiconductor components. The components only function according to specification within the voltage ranges supplied by the manufacturer's data sheets. The circuit behaviour is indeterminate outside of this range. After the voltage returns to this specified range, the component once again functions according to specification. Once the destructive voltage limits are exceeded, permanent component damage can be expected.

1.2.3 Sources of Interference Generated by Transmission Cables for Digital Circuits

Interfering signals may be picked-up by the connection cables and transferred to the circuits of an electronic system. The cable and circuit acting in combination may also lead to erroneous circuit functions.

External capacitive coupled interference pulse

External interference pulses originate outside of and are coupled into an electronic system. For "short" cable lengths, these are mainly capacitively coupled into the circuit (see section 3.2). The term "short" cable refers to all cables having a signal propagation time much smaller than the rise or fall times (distance between the edges of the pulse slope) of the transmission signal. (Approx. values: The frequency dependent signal propagation speed corresponds to approx. 0.6 times the speed of light for high quality cables and approx. 0.3 times for standard cables.) A discrete quantity of interference energy is required to cross the dynamic signal-to-noise threshold of a circuit. As the typical input resistance of a circuit is large, only the magnitude of the mutual capacitance and the output resistance of the signal transmitter remain important. These two build a resistance-capacitance element, whose time constant determines the duration of and thereby the amount of energy introduced by the interfering impulse. When using the same cable type, circuit families with a smaller output resistance are more resistant to interference impulses.

The following example aids in the estimation of a so-called "short" cable.

Example

An impulse having a transmission frequency of 10 MHz is to be transmitted. Electric waves propagate along the designated copper cable at approx. 0.6 times the speed of light. The edges of the impulse's slope are separated by a width of 10 ns. If the signal's slope has a width of 10 ns, a maximum permissible signal propagation time t_L of 0.5 ns can be calculated for a "short" cable. The calculation results in

$$L_{max} = 0.6 \cdot c \cdot t_L$$

a cable length L_{max} of 90 mm.

Transmission distances of this length are generally not bridged with insulated cables. Thus, most insulated cables have the characteristics of "long" cables (section 1.2.4).

Resident capacitive coupled interference pulse

Within an electronic system, mutual capacitance may cause spurious signals to be transmitted between neighboring cables. In contrast to external capacitive coupled interference pulses, the non-negligible small internal resistance of the interfering source (the output resistance of the transmitter) must be considered. The magnitudes of the input and output resistances and the magnitude of the mutual capacitance determine the amplitude and amount of energy for the capacitive coupled interference impulse. To ensure that the energy required to exceed the dynamic signal-to-noise ratio is minimized at the input, not only should the mutual capacitance between the cables be reduced, but also the output resistance of the circuit. These values change with the switching conditions of the circuit. Whether or not an interference impulse exceeds the dynamic signal-to-noise ratio of the circuit, consequently depends upon the voltage level on the transmission cable. TTL-circuits, for example, are less sensitive to such disturbances at the low level than at the high level.

1.2.4 Electrical Waves at Junctions

A wave arriving at a junction between two mediums having different characteristic impedances will be either partially or totally reflected. The numerous reflected waves at the end of a cable overlap the waves of the actual signal. The path of the resultant wave at the input of the circuit no longer corresponds to the original signal, and the circuit reacts unpredictably.

For "short" cables, the reflections fade during the signal switching time due to the short signal propagation times within the cable. This results in negligible deviations in the signal's rise and fall times. Therefore, the propagation effects for "short" cables do not require additional consideration.

"Long" cables have relatively long signal propagation times in comparison to the signal switching time. The propagation times cannot be altered afterwards, as they are dependent only upon the dielectric coefficient ε_r of the conductor's environment and the relative permeability μ_r of the conductor.

If a "long" cable is not terminated at the ends by its characteristic impedance, an electrical wave (e. g. an impulse) will be partially reflected and will travel at its propagation speed back and forth between the two cable ends until its energy dissipates. This implies that with each length traveled, the amplitude of the wave shrinks proportionally to the cable's attenuation factor. During this time, the signal at the output of the transmitter (i.e. the input of the cable) may change. It is easy to imagine that an operating signal superimposed with several reflected waves produces a different value at the receiver input than originally transmitted. The value is dependent upon both the amplitude and phase position of the signal. The voltage may both positively and negatively overshoot at both cable ends, which may result in erratic behaviour at the receiving circuit module. This behaviour is especially noticeable in circuits having low output and high input resistances, which is exactly the case for typical circuits.

This reflection condition becomes extreme for very small output and very large input resistances. The wave is often reflected and is weakened only slightly at the cable ends. Continuing oscillations result along the cable. This can only be controlled by preventative circuit design measures.

1.3 Modes of Transmission

Symmetrical and unsymmetrical transmission principles differ in regard to noise immunity. It is important that the entire transmission path (the transmitter output stage, transmission cable and receiver input stage) is constructed entirely according to one principle. Customarily, more than one transmission principle is employed, for example: the cable has a balanced construction in contrast to unbalanced transmitter and receiver stages. Improvements in the electromagnetic compatibility may in the latter case only be partially realized.

Within this book, the German term *unsymmetrisch* has been translated with *unsymmetrical* and *symmetrisch* with *symmetrical,* as well as *asymmetrisch* with *asymmetrical.*

1.3.1 Unsymmetrical Transmission

Unsymmetrical transmission results when two conductors having different potentials (voltages) are used to transmit a signal within a system also having an unbalanced reference potential. This is usually the earthed reference conductor for DC voltage supply.

Diagram 1.4 depicts two examples of unsymmetrical signal transmission, whereby for diagram 1.4a the signal is generated with a switch in the transmitter and for diagram 1.4b the signal is introduced externally in the transmitter output stage.

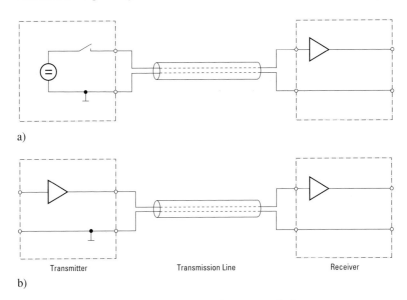

a)

b)

Transmitter   Transmission Line   Receiver

Diagram 1.4   Examples of unsymmetric signal transmission

21

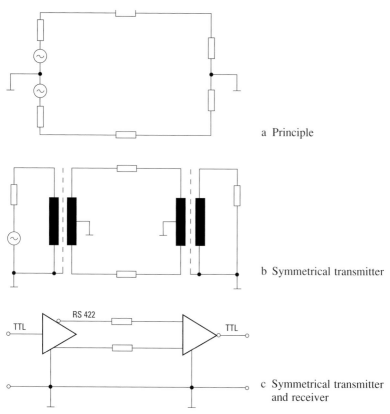

a Principle

b Symmetrical transmitter

c Symmetrical transmitter
and receiver

Diagram 1.5   Examples of symmetrical signal transmission

1.3.2 Symmetrical Transmission

In contrast to unsymmetrical transmission, the potentials of the outgoing and returning signal cables are for practical purposes of equal magnitude *and* are balanced with respect to the reference potential. This transmission principle offers better reliability, but may require more complicated circuit and cabling design. Diagram 1.5a depicts the symmetrical transmission principle. Here the impedances of both the outgoing and returning cable are identical (balanced) at the signal source, at the receiver and within the cable. In diagram 1.5b the transmitter and receiver are galvanically separated. Diagram 1.5c shows a balanced liner driver and line receiver.

The advantages of symmetrical transmission are clarified in the following comparison.

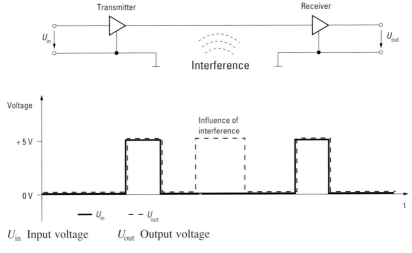

U_{in} Input voltage U_{out} Output voltage

Diagram 1.6   Voltage level for an disturbed, unsymmetrical transmission

Diagram 1.6 demonstrates the signal progression of unsymmetrical transmission. The interfering signal is simply portrayed as a rectangular signal. It is introduced into the transmission cable and appears at the input of the unbalanced receiver, where it is further propagated.

Diagram 1.7 illustrates the interference effects on a symmetrical transmission path. The interfering signal is introduced by a balanced transmission cable in both conductors with identical magnitudes and polarities. The voltage difference between the two conductors is thereby not altered. Thus, the output of the balanced differential op-amp remains unaffected by the interference signal.

1.3.3 Crosstalk during Transmission

During transmission, unwanted coupling effects between neighboring cables, the so-called *crosstalk,* may occur. Two sub-categories of crosstalk are half-duplex and duplex interference.

Half-duplex interference

During half-duplex interference, the operating signals (as well as impulses) transmitted along neighboring cables are temporally staggered, i.e. they do not overlap. The signal in one cable may induce an electrical wave in the other cable causing interference. For further information, the standardized parameter *near end crosstalk loss* is defined in section 6.5.

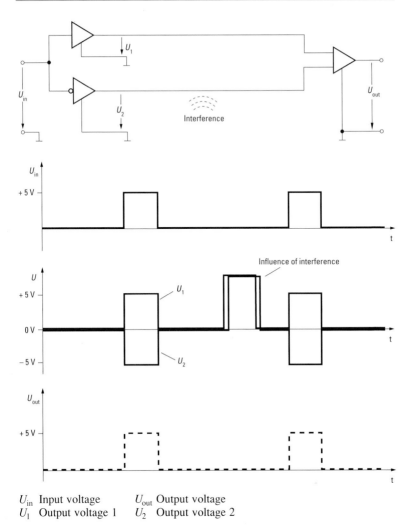

U_{in} Input voltage $\quad U_{out}$ Output voltage
U_1 Output voltage 1 $\quad U_2$ Output voltage 2

Diagram 1.7 Voltage level for symmetrical transmission under the influence of interference

Two different operating cases exist:

1. Both transmitters are placed on the same side of two cables laid in parallel.
2. Both transmitters are placed at opposite ends.

The first case is referred to as having "parallel" running transmission cables, and the second, as having "non-parallel" running transmission cables.

Half-duplex interference along "parallel" running transmission cables can only lead to interference for extremely short signal rise/fall times combined with very long cable lengths. The effects of crosstalk are more critical for "non-parallel" running transmission cables linked with low-resistance transmitter outputs and high-resistance receiver inputs. Hereby, overly large voltage peaks and erroneous signals due to oscillations may arise at the ends of the disturbed cable. The excessive voltages are proportional to the conductor's capacitance with respect to earth. If this capacitance is large, the voltage overshoots will become even larger.

Duplex interference

Duplex interference arises when the signals transmitted by both cables temporally coincide. In this situation, the "non-parallel" running cables are also more sensitive to interference. The impulse rise/fall times of the operating signal lead to interfering voltage overshoots on the neighboring cable. A reduction in the cable's inductivity minimizes the duplex interference between the conductors of the cable. This can be achieved, for example, by including an additional earthed conductor within the same cable.

Symmetrical transmission requires balanced cables. Balanced cables possess significantly more favorable half-duplex crosstalk values than unbalanced cables. Furthermore, symmetrical transmission demonstrates far better electromagnetic compatibility (EMC) than unsymmetrical transmission as will be shown in chapter 6.

1.4 Earth and Signal Ground

In common practice, the earth and signal ground conductors are not differentiated. One cable is employed for both functions. These two conductors should, however, be strongly distinguished for signal transmission applications. Separate cabling systems should be constructed to facilitate noise immunity.

For purely functional reasons, a circuit does not require an earth connection. An earth is not necessary to process the sequences of a circuit function.

1.4.1 Earth

An earth (safety ground) is necessary to safely operate electrical equipment. Earthing conductors are not galvanically connected to the other circuit conductors. They only conduct current in the case of error.

In practice, the earthing conductor does not have ground potential at every point along its length. This may be caused by a mains interference suppression filter, for example, which does not allow the insignificant arrester discharge current to flow through the earthing conductor. The potential gradient caused by the arrester discharge current often lead to common-mode voltages within the ground loop and thus, consequently, to interference voltages in the circuit (see chapter 3).

1.4.2 Signal Ground

The signal ground (circuit common) serves as a common reference conductor within an electric circuit. Reference conductors carry operating current. The signal ground can but does not have to remain at ground potential.

In order to prevent the presence of undesirable potential differences, electronic systems are earthed via its reference conductor (signal ground). If the extension and thus the length of the reference conductor is longer than the wave of the transmission signal, then the system should be earthed at several points. This is known as multi-point grounding. Single-point grounding can be bridged by capacitive coupling. If the system extension is shorter than one seventh of the wavelength of the transmission signal, single-point grounding should be implicitly employed. If the wavelength lies between the two values, then single-point grounding with the use of possible restrictions is preferable.

If the signal ground is earthed at several different locations, ground loops result. Due to their varying ground potentials, compensation currents flow along these signal grounds. These currents may result in differential interference voltages within the circuit. The exception being symmetrical transmission (see chapter 4).

If the length of the reference conductor approaches that of the wavelength, its impedance rises boundlessly, and the signal ground only remains effective through parasitic capacitance and magnetic coupling.

The further away a grounding point is, the greater the actual potential of an earthed reference conductor deviates from the earth potential. This is caused by voltage losses along the reference conductor due to flowing operational current.

The following are interchangeable terms having the same definition for "earth" and "signal ground".

Earth:

Earth ground, protective earth, fault protection, ground earth, equipment ground, safety ground and so forth.

Signal ground:

Signal ground, signal reference, control common, circuit common, neutral, 0-V-bus and so forth.

1.5 Equipment Interfaces

1.5.1 Serial Interfaces

Serial bus systems are largely based on standardized interfaces. They allow one to fall back on known, proven hard- and software components for transmission technology. Within serial bus systems, data is bidirectionally transmitted over a two-conductor cable. Table 1.3 contains details concerning standardized serial interfaces. The information presented in the table is without guarantee.

Table 1.3   Serial interfaces

Desig-nation	Type	Maximum driver output voltage V	Maximum data rate kbit/s	Maximum cable length* m	Maximum number of participants	Driver load Ω
V.24/V.28	unsymmetric	25	20	15 (50)	1 R / 1 D	3...7k
RS-232-C	compatible with V.24/V.28					
V.10/X.26	unsymmetric	6	100	1200 (4000)	10 R / 1 D	450
RS-423-A	compatible with V.10/X.26					
V.11/X.27	symmetric	−0.25...6	10000	1200 (4000)	10 R / 1 D	100
RS-422-A	compatible with V.11/X.27					
RS-485	symmetric, contains RS-422-A	−7...12	10000	1200 (4000)	32 R / 32 D	60
TTY 20 mA		20 mA ± 20 %	20	1000	1 R / 1 D	

* the permissible cable length often depends upon the chosen data rate
R Receiver
D Driver
( ) under favorable EMC conditions

1.5.2 Parallel Interfaces

Parallel Interfaces facilitate transmission with higher data rates than serial interfaces, since the bits of one or more bytes are simultaneously transmitted in parallel. Every bit queued in parallel needs a separate conductor in the cable. The maximum achievable transmission length is significantly shorter than for serial transmission. Therefore, parallel interfaces have almost only gained acceptance in equipment technology. There are only a few parallel interfaces having broader acceptance (e.g. the non-standardized CENTRONICS interface). Table 1.4 contains details to several known parallel interfaces and bus systems used in building systems technology. The information in the table is provided without guarantee.

Besides data cables, clocking, feedback, signal, control, auxiliary and earthing cables are connected via parallel interfaces. The system designer has the freedom to choose which cables are applicable for the system in question. Therefore, general and valid statements concerning the number of necessary conductors cannot be made.

Table 1.4   Parallel interfaces/Bus systems

Designation	Transmission mode	Maximum driver output voltage V	Number of conductors per cable	Maximum data rate kbit/s	Maximum cable length m	Maximum number of participants
IEC-625	unsymmetric	5	25	500	20	15 R / 1 D
IEE-488	unsymmetric	5 (TTL)	24	500	20	15 R / 1 D
CENTRONICS	unsymmetric	12	18	240	approx. 2	1 R / 1 D

R Receiver   D Driver

2 Bus Systems in Building Installations and Industry

Bus systems are increasingly replacing conventional cabling for signal and data transmission. They are less expensive and less prone to interference than previous point-to-point cabling structures. Bus technology has principally developed through office and industrial applications. Bus technology is currently adopted either directly or in a modified form within building installations. The following applications, though drawn from industry, are just as valid for building systems technology (often termed "intelligent buildings").

2.1 Cable Management Systems

The predominating electronic systems in industry (e.g. control and instrumentation systems) primarily employ star-shaped cable structures. This means that all peripheral (switches, sensors, actuators, etc.) are connected individually, via a separate cable, to a central unit (e.g. programmable logic controller PLC). Signals are bidirectionally transmitted between the central unit and peripherals (terminals). Diagram 2.1 shows a basic star-shaped cable structure in which each peripheral unit is connected via a separate cable to the central unit.

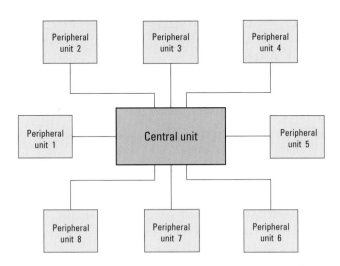

Diagram 2.1   Basic star-shaped cabling structure

The amount of cabling required, even for mid-sized systems, is substantial and increases rapidly for more physically extensive systems. It is important to consider the extra effort and expense necessary for connectors, thermal and mechanical protection, cable tracks and fastening systems. Wear-out due to friction and abrasion of the free-flexing signal cables employed in industrial environments often leads to expensive stand-stills in processing equipment.

Over the past years, one trend is to transfer commercial and administrative data processing to the manufacturing floor and vice versa. Data processing completed by machines or industrial equipment are blending with the organisational controlling and accounting systems of corporate management. A prime example is CIM (Computer Integrated Manufacturing). A computer system designated for this purpose allows information to be transferred across all hierarchical levels within a corporation. Thereby, information may be exchanged at any time between all participants. Such a system naturally demands that all computer, control and data capturing systems are linked via the appropriate cables. The cabling expenditures for a star-shaped or even a meshed structure would be prohibitive.

2.2 Bus Cabling Structures

In order to reduce cable expenditures, all terminals are connected with a single cable either individually or in groups. Likewise, the central unit may be linked to this cable (bus).

For signal transmission along the bus, the following applies:

1. Transmission along the bus is bi-directional.

2. As all terminals (peripheral units) are connected to one cable, all signals will reach all terminals.

3. To ensure that only the intended terminal receives the destined information, the transmitting terminal must include a recipient address with the signal. This is known as a telegram.

4. Only one telegram may be transmitted along the bus at any time. Otherwise, the messages merge into a confused jumble.

5. If several telegrams simultaneously appear, their sequential transmission must be regulated. This must ensue rapidly to avoid unacceptably long waiting periods and possible interference during serial data transmission. Telegram transfer combined with short waiting periods assumes a very fast transmission method.

Since the information transmission process is comparable to passenger bus services, the term "bus system" was adopted. A passenger (telegram)

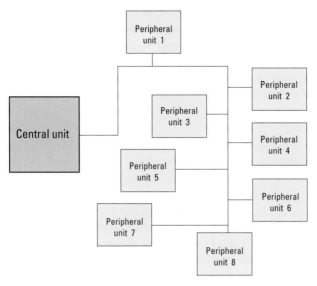

Diagram 2.2   Basic bus cabling structure

can board the bus (transmission line) at any bus stop (transmitting terminal) and can travel to its chosen destination (receiving terminal). The basic bus structure for cabling is depicted in diagram 2.2.

Various forms of serial bus systems presently exist. Either stranded pairs, coaxial or two-conductor cables are used as the transmission medium. Mineral or synthetic fiber optic cables have recently been employed in special applications. This book focuses on copper bus line systems. Fiber optic technology would explode the scope of this book and is reserved for specialized technical literature.

For physical reasons, only serial bus systems are suitable for fast signal transmission over long distances (> 5 m). Parallel bus systems are almost exclusively used within equipment housing (e. g. PCs, PLCs, measurement devices, etc.). They are largely employed as "data acquisition channels". Since this book concentrates on cable connections between equipment, parallel bus systems will not be further described.

2.3 Electrical Bus Couplers

A bus coupler electrically connects a peripheral to the bus line. They consist of the digital circuits described in chapter 1. The input and output circuits on the bus line do not necessarily require a circuit technology

31

identical to the terminal. To achieve high interference immunity, circuit technologies having a higher signal-to-noise ratio are preferred. Thus, their disadvantages are occasionally acceptable.

2.4 Bus Systems

Bus systems extend from the low and well into the high frequency ranges. Consequently, a system's cabling complexity depends upon both transmission quality and electromagnetic compatibility (EMC). The development and application of a "standard" cable fulfilling all electro-mechanical requirements is technically and economically hardly conceivable.

Most bus systems are based on standards. Some carry a specific trademark and often either surpass or deviate from the basic stipulations of the governing standard. Table 2.1 lists the bus systems typically used in industry.

The standards establish requirements for the transmission cables generally in the form of minimum profiles. This may include the determination of cable transmission parameters as well as general recommendations such as for "shielded twisted-pair cables". Rarely, do they contain complete cable specifications. Special features concerning flexibility, halogen-content or burning behaviour are not defined. Even the design measures necessary to ensure electromagnetic compatibility are not described. The systems designer or user must determine these details. Consequently, many opportunities to engineer the data transmission cables along with their installation and connection technologies remain open.

Table 2.1 includes the ISDN communication system, although, it is not a bus system. Since the routing of communication cables within a building often share the same pathways as those of data transmission cables, it seems practical to use the conductors of one cable for the two different systems. Cost factors suggest employing a single type of cable for various systems including ISDN.

All bus systems cannot be comprehensively explained at this point. The majority of systems are based on the previously mentioned serial interfaces or bus standards. Occasionally, they are tailored to special applications or modified for environmental conditions. Furthermore, standards exist in which only the cable structures, including some of the above mentioned bus systems, are specified. These describe in detail the transmission medium, the electromechanical system parameters and the entire cabling structure (including electrical connections). Several examples follow:

Table 2.1 Examples for common industrial bus systems

Designation	Reference standard	Transmission medium	Maximum transmission rate and frequency	Maximum cable length m	Transmission mode	Product names or Comments
Ethernet						
10 Base 2	IEEE 802.3	coaxial	10 Mbit/s	200	unsymm.	Cheapernet
10 Broad 36		coaxial	10 Mbit/s	3800		
10 Base 5		coaxial	10 Mbit/s	500	unsymm.	Yellow Cable
10 Base T		UTP, quad	10 Mbit/s	100		
10 Base F		Fiber Optics, 850 nm	10 Mbit/s			
	IEEE 802.3	triaxial	10 MHz	500		SINEC H1
PROFIBUS	DIN 19245	STP, pair	9.6... 500 kbit/s	1200... 200	symm.	SINEC L2 SUCONET-P INTERBUS-P
	+ Token Passing		9.6... 1500 kbit/s	5000... 200	symm.	SINEC L2
	DIN 19245-3	STP, pair	... 1.5 MHz	200	symm.	SINEC L2
INTERBUS S	DIN 19258 (RS 485)	Li-YCY $3\times2\times0.25$ mm²	500 kbit/s	13000	symm.	VARICON
Bitbus	IEEE 1118 (RS 485)	UTP $3\times2\times*$ STP $3\times2\times*$	375...63 kBaud 269...77 kBaud 375...62.5 kbit/s	300...1200 300...1200	symm.	INTERBUS-C
EIBUS	DIN VDE 0829	YCYM $2\times2\times0.8$ mm	9.6 kbit/s	1000	symm.	
ISDN Based connection		$2\times2\times0.6$ mm	144 kbit/s avg.			
Train/ Rail-bus	E DIN 43322 RS 485	STP 2×0.5 mm²		200	symm.	
ISU-BUS	ISO 8802.4	STP $2\times2\times0.5$ mm²	256 kBaud	50 km		for mining

STP Shielded Twisted Pair   UTP Unshielded Twisted Pair   Fiber Optics = Mineral glass fiber optic cable   S-shielded and U-unshielded refer to the individual stranding elements and do not imply an overall shield for the cable.
* not established

Working paper from committee No. 715.0.3 WG3 N76: "Guideline for the Planning and Installation of Information Technology Cable Systems Serving Token Ring Stations According to ISO 8802.5"

prEN 50098-1, Part 1: "Informationstechnische Verkabelung von Gebäudekomplexen; Aufbau and Dimensionierung von Kabelnetzen für den ISDN-Basisanschluß"

EIA/TIA-568: "Commercial Building Telecommunications Wiring Standard"

EIA/TIA TSB-36: "Technical Systems Bulletin Additional Cable Specifications for Unshielded Twisted Pair Cables"

NEMA: "User Guide to Product Specifications for Electrical Building Wire and Cable"

Today, cables are no longer exclusively used to transmit data within electronic equipment or systems. They are also used to network highly diverse systems within a corporation. The resulting networks are known as LANs (Local Area Network). The term "local" signifies a limited area of coverage and that the system is legally at the sole disposal of the operator.

2.5 Protection Against Dangerous Currents within Bus Systems

Extensive bus networks may place higher demands on the protection measures against dangerous currents. The fundamental principles for a specific system such as a bus system for building installations (e. g. European Installation Bus EIB) must be considered.

Protection measures for the EIB-system

Due to a high electromagnetic compatibility, the EIB-system strongly suggests laying bus cables in close proximity to power cables. The bus system is operated at a safety extra low voltage SELV (recently known as "protective extra low voltage" according to DIN VDE 0100 Part 410/11.83 4.1). Consequently, according to DIN VDE 0100 Part 410, the active components within both standard voltage or protective extra low voltage circuits must be safely separated. This safety measure can be achieved in bus cables by adding supplementary insulation. For this purpose, the EIB-cable draws upon the additional insulating properties of the outer jacket. The dielectric strength of the jacket is tested at 2.5 and 4 kV. If a voltage surge is transferred from the low voltage network onto the bus network, unconditional shock protection against accidental contact with voltage carrying components is provided. Herewith, the continued functionality of the entire EIB-network is also guaranteed. The propagation of low voltages along the bus line remains restricted, since the line coupler is galvanically separated from the bus line, and since all components are implemented according to the SELV guidelines. Therefore, backward voltage surges running from the low voltage carrying conductors of the bus line into neighboring metallic objects do not occur. The additional insulating properties of the bus cable's outer jacket aid in ensuring such safety measures.

The data cables within the EIB-system are an important component of the safety concept, which provides shock protection in the event of system failure. The dielectric strength of the outer jacket (insulating property) is a safety feature which must be fulfilled by the data cable along with the usual signal transmission requirements.

The importance of these safety measures is better illustrated by the following calculation:

An EIB-system having the maximum number of subscriber extensions ($13 \times 15 = 195$ bus lines each with a max. cable length of 1000 m) is installed in a building used for administrative functions. Using a conservative estimate of 0.5 m/m^2 for the amount of cables per floor, a maximally extended EIB-system would cover a surface area of 400,000 m^2. Using a max. number of 64 terminals per bus line, up to approx. 12,480 terminals may be connected to the entire system.

Due to the extensive coverage and the large number of equipment connections, the potential danger in the event of failure would affect many people. Protection against dangerous electric currents is provided by the above mentioned safety precautions.

2.6 Cable Infrastructures within Buildings

Cable infrastructures for information technology within building installations hardly differ from those used for industrial applications. Cabling for buildings must, however, be planned to last over a considerably longer period of time. The anticipated lifetime for machines and industrial equipment is between 5 to 15 years. Over this duration, the built-in electronic systems rarely change. Therefore, the cabling structures generally undergo only slight adaptations to minor modifications in the electronic systems.

The concepts for flexible cabling infrastructures lasting far into the future are radically different. They must meet all future technical demands without requiring complex changes or adaptations until the building is either torn down or undergoes a comprehensive renovation. This implies that the cabling structures must be suitable for all future electronic systems which may not even exist in the design stage at the time of the building's erection. Rewiring a building is an extremely expensive endeavor. Furthermore, the various electronic systems of the future must share a common network to reduce associated costs as well as space and weight allocations. This goal sets technical limitations. At least the most essential systems should employ a common type of transmission cable, despite the variety of cabling structures involved. This is especially valid for:

▷ building installation technologies (intelligent building systems)

▷ building control engineering

▷ building management systems

▷ computer systems and computer links

▷ security systems (e.g. intruder alarm systems)

▷ telecommunications and television (including closed-circuit TV).

The ISO/IEC DIS 11801 standard "Information technology Generic Cabling for Customer Premises Cabling" specifies a complete network-neutral, future-oriented cabling structure for buildings. Only copper cables are recommended for the horizontal cabling of individual floors. Copper cables are sufficient for star-shaped cable structures. Since the cable end of each data port must offer a variety of services, this standard also defines appropriate requirements for the corresponding connection cables. The services described above are divided into classes according to their maximum permissible transmission frequencies. The cables are further grouped into cable categories. A specific set of requirements is defined for each cable type within a cable category (section 6.5). Each class also restricts the maximum permissible length of the cable.

Table 2.2 lists the maximum permissible cable lengths for balanced 100-Ω-LAN cables recommended by this standard. The applications for the cabling systems are assigned to classes "A" to "D". Class "A" has the lowest rank. The lengths given in table 2.2 are valid only, if the near-end cross-talk loss (see section 6.5) and the wave attenuation requirements for the individual classes are met, and only, if the specified components are used.

ISO/IEC DIS 11801 merely defines transmission characteristics of the cables, but does not contain any cable specifications. Further electrical and mechanical parameters may either be freely chosen to suit the application or may be determined by the governing national cable standard.

Table 2.2 Permissible cable lengths according to ISO/IEC DIS 11801

Class	Maximum transmission frequency MHz	Maximum permissible cable length		
		Cable Category 3 m	Cable Category 4 m	Cable Category 5 m
A	0.1	2000	3000	3000
B	1	500	600	700
C	16	100*	150**	160**
D	100	–	–	100*

* The 100 m length includes: max. 10 m for flexible jumper cables and work station or equipment connection cords (for further information see ISO/IEC DIS 11801)
** The specific application standards should be referenced for lengths over 100 m.

The cable's shield may be constructed according to the manufacturer's or customer's wishes.

The implications for the user include:

The assignment of a cable to a "cable category" depends only upon its transmission properties. All other characteristics must be either separately agreed upon or taken from the agreed upon national cable standard. Membership of the same category is, therefore, not the only prerequisite for a cost comparison between different cable types. The cables must also offer comparable features which are not defined by the standard (section 6.2).

Data ports

Cable structures, which anticipate future demands on building installations according to ISO/IEC DIS 11801, call for at least two data ports (connection boxes) per work station for applications within the realms of "Class A" to "Class D":

Port 1 Interconnection to a cable having at least "cable category 3"

Port 2 Interconnection to a cable having at least "cable category 5"

Table 2.3 Networks and services for structured, generic cabling for building installations (customer premises) (Examples)

Class	Network	Standard	No. of Pairs	Remarks
Class A 0.1 MHz	PBX X.21/V.11 S_0-Bus (extd) S_0-point-to-point	National regulations CCITT Recs.X.21/V.11 CCITT Rec.I.430 CCITT Rec.I.430	1...4 2 2(...4) 2(...4)	ISDN Basic Access ISDN Basis Access
Class B 1 MHz	s_1/s_2 CSMA/CD 10 Base5	CCITT Rec. I.431. ISO/IEC 8802.3	2 2	ISDN Primary Acc. Ethernet
Class C 16 MHz	CSMA/ CD 10 Base-T Token Ring 4 Mbit/s Token Ring 16 Mbit/s	ISO/IEC 8802.3 DAM 9 ISO/IEC 8802.5 ISO/IEC 8802.5/ DAD 1	2 2 2	Ethernet
Class D 100 MHz	Token Ring 16 Mbit/s ATM (TP) TP-PMD	ISO/IEC 8802.5/DAD 1 CCITT and ATM-Forum ISO/IEC	2 2 2	B-ISDN Twisted Pair DDI

( ) Reserve conductors for additional voltage supply

The largest transmission reserves are naturally acquired by connecting two data ports using cables from "cable category 5". The increased cable prices are somewhat compensated by the reduced effort and the cost saved during installation and management of the network. When considering the advantages of a future-oriented building system over decades, the initial financial expenditures no longer carry the same weight. Diagram 2.3 depicts a generic cabling structure for building installations.

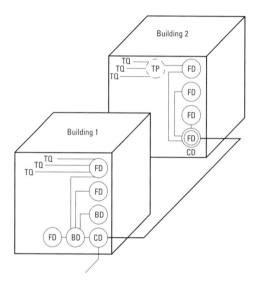

TQ Telecommunications-
    port
TP Transition point
FD Floor distribution
CD Communal
    distribution
BD Building distribution

Diagram 2.3
Generic, structured cabling for building installations

3 Mechanisms of Electromagnetic Interference

The quality of information transmitted over cables depends not only upon the transmission efficiency of the cables, but also upon the noise immunity of the entire system to electromagnetic influences from the surrounding environment. Conversely, the transmission system also influences its environment by generating electromagnetic fields. This correlation is known as electromagnetic compatibility EMC (see chapter 5).

Without exception, all electrical equipment and systems are as of 3rd May 1989 subject without restriction to the guidelines 89/336/EWG. These guidelines are recommended by the European Union in order to unify the legal regulations of the member nations concerning electromagnetic compatibility. In Germany, the EMC guideline was converted into a law on 13th November 1992 entitled "The Electromagnetic Compatibility of Equipment" (EMVG). The product characteristics to be demonstrated are the appropriate susceptibility and the allowable emission. These features are based on harmonized standards from CENELEC, which in turn were partially originated in CISPR and IEC publications.

In order to specifically plan the electromagnetic compatibility of a system, the interfering environment *(noise source),* the coupling mechanisms and the receptivity or sensitivity of the endangered system *(noise sink)* must be known.

Disturbance variables (voltages) reach the interference sink by various means. When transferred via cables and passive components, it is referred to as a conducted transmission of the disturbance variable. In other cases, coupling or radiation is present.

As long as the wavelength is large in comparison to the size of the interference source, the electromagnetic influences spread predominantly via galvanic (cable-related), capacitive or inductive coupling. If the wavelength and dimensions of the voltage carrying element are comparable, then radiative interference takes effect.

Whilst interference sources and sinks can be easily measured, the coupling mechanisms usually arise unexpectedly as parasitic effects. Fundamentally, one discerns between four coupling mechanisms:

▷ galvanic coupling

▷ capacitive coupling

▷ inductive coupling

▷ radiative coupling.

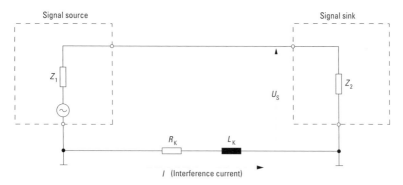

I   Interference current
R_K Resistance of the reference potential
L_K Inductance of the reference potential
U_S Voltage across Z_2 due to the interference current
Z_1 Internal resistance of the signal source
Z_2 Internal resistance of the signal sink

Diagram 3.1   Functional principle of galvanic coupling

3.1 Galvanic Coupling

A galvanic coupling (diagram 3.1) may occur, when several electric circuits partially or completely share the same conductor. The mutual voltage drop caused by the combined currents along this section of conductor, results in differential interference voltages within every circuit employing unsymmetric transmission. Symmetric transmission does not induce differential interference voltages in this manner, as each circuit has its own outgoing and returning conductor. Galvanic coupling, always requiring an electric conductor, belongs to the cable-related (conducted) interferences.

3.2 Capacitive Coupling

Capacitive coupling arises due to the dielectric between conductors having different potentials. The alternating field caused by the AC voltage of the interfering conductor drives an interfering current through the disturbed circuit via mutual capacitance and back to the interference source via either the signal ground or earth (diagram 3.2). The interfering current generates additional voltage drops within the disturbed circuit. These overlap the operating signal and may cause distortion at the receiver.

The electric field stands perpendicular to the conductor axis. It can exist without a magnetic field. If the wavelength λ is smaller than seven times

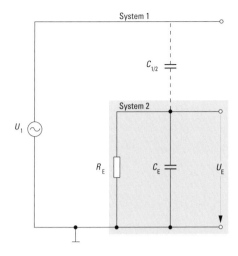

U_1 System voltage 1
U_E System voltage 2
R_E Replacement resistance for the internal resistance of the transmitter and receiver in system 2
C_E Replacement capacitance for the capacitance of the transmitter and receiver in system 2
$C_{1/2}$ Parasitic capacitance between system 1 and system 2

Diagram 3.2   Functional principle of capacitive coupling

the length of the cable L ($\lambda < 7\,L$), a magnetic field will always be generated in conjunction with the electric field. With relatively little effort, shielded cables easily weaken electric fields.

At frequencies below 30 MHz, the intensity of inductive coupling is noticeably greater than that of capacitive coupling. Thus, only the magnetic field of a current becomes relevant, if one actually flows.

3.3 Inductive Coupling

An alternating magnetic field, which stands perpendicular to the surface enclosed by the conductor loop of a circuit (outgoing and returning conductors), induces a normal-mode interference voltage within this circuit. The magnitude of the differential interference voltage introduced is proportional, among other things, to the surface area encompassed by both conductors of the circuit and proportional to the strength of the alternating disturbance field (diagram 3.3).

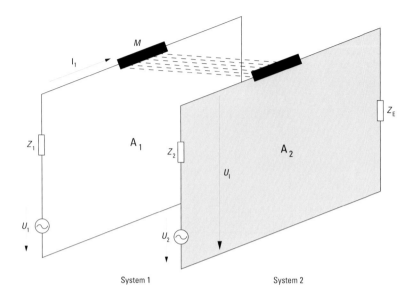

U_1	Voltage in system 1
U_2	Voltage in system 2
U_I	Interference voltage in system 2
M	Mutual inductivity
A_1	Area of the conductor loop in system 1
A_2	Area of the conductor loop in system 2
Z_1	Impedance of the transmitter in system 1
Z_2	Impedance of the transmitter in system 2
I_1	Current in system 1
Z_E	Impedance of the receiver in system 2

Diagram 3.3   Functional principle of inductive coupling

Inductive coupling cannot be entirely eliminated by connecting the cable's shield to earth. A shielded cable is only effective against inductive coupling, when *both ends* of the shield are earthed forming a closed circuit. At higher frequencies, parasitic capacitance takes effect. Hereby, current may begin to flow along a non-earthed or single-sided earthed shield. Such a poorly earthed shield is only partially effective.

When twisted pair conductors are present in an alternating field, the induced partial voltages are neutralized within the individual loops of each twist. However, this cancellation occurs in so far as the sizes of the loops remain uniform. The remaining differential interference voltages induce normal-mode interference voltages between the ends of a conductor pair (see section 9.1).

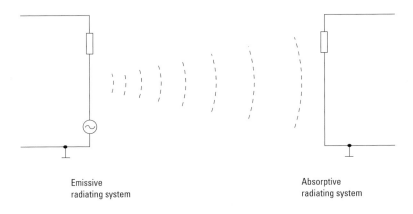

Emissive
radiating system

Absorptive
radiating system

Diagram 3.4 Functional principle of radiative coupling

Principally:

Effective shielding against magnetic fields can only be realized through great effort.

3.4 Radiative Coupling

As the length of the interfering cable approaches or exceeds the length of the interfering wave (> approx. $\lambda/7$), electromagnetic waves are emitted. This implies that there is always an electric and a magnetic field present. The magnetic field always stands perpendicular to the electric field. They are coupled with one another through the wave impedance of the surrounding environment. In comparison to capacitive coupling, the electric field components in parallel to the cable are of importance. Diagram 3.4 portrays both emissive and absorptive radiating systems.

The coupling effect is difficult to calculate exactly. Approximation methods are based on antenna theory.

3.4.1 Near, Far and Transition Zones

These terms are customary in antenna technology.

Far zone

Within the far zone of an antenna, an electromagnetic wave (H-field and E-field) is always present. All field components are of equal value. Far zones begin where the separation r between the transmitting and receiving antennae is greater than three times the wavelength ($r > 3\ \lambda$). A realistic value for the characteristic impedance of a field in free-air is 377 Ω.

Near zone

Within the near zone ($r < 0.1\ \lambda$) of an antenna, the antenna itself determines whether or not an electric or magnetic field is formed. The magnetic field dominates in the near zone of an interference source, if the system has a low-resistance or low-inductance. For example, inhomogeneity in the shield (holes) always produces a magnetic field (*H*-field) on the opposite transmitter. High-resistance or high-inductance systems are characterized by electric fields. The complex characteristic impedance of the field is correspondingly capacitive or inductive.

Typical cable lengths for building installations or industrial applications range up to 10 meters. A typical edge-length for the equipment housing is roughly 1 m. Using these dimensions, near zones are present at interfering frequencies up to approx. 30 MHz. Far zones exist at higher frequencies. With respect to electromagnetic compatibility, shielding measures against electric *and* magnetic fields should always be taken in the far zone of an unknown interference source. The attenuation of a shield is far more effective against electric than magnetic fields. Within the near zone, protection against magnetic fields usually suffices.

Transition zone

The region between near and far zones is called the transition zone ($0.1\ \lambda < r < 3\ \lambda$).

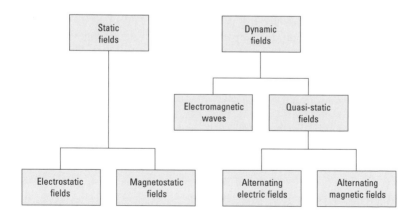

Diagram 3.5   Field types

3.4.2 Field Types

One differentiates between *static* and *dynamic fields*. The latter is further sub-divided into *quasi-static fields* and *electromagnetic waves*. These classifications are illustrated in diagram 3.5.

Static fields do not change during the period of observation. The time dependency of a quasi-static field is so small, that an alteration in the region under consideration (e.g. in the near zone of an antenna) is simultaneously noticeable everywhere within the region. In comparison, field conditions for *electromagnetic waves* during the observation period change so quickly that both its local and temporal development must be considered.

3.5 Transmission Methods of Disturbance Variables

Disturbance variables (voltages) reach the interference sink by various means. When transferred via cables and passive components, it is referred to as a conducted transmission of the disturbance variable. In other cases, coupling or radiation effects are present.

As long as the wavelength is large in comparison to the size of the interference source, the electromagnetic influences spread predominantly via galvanic (cable-related), capacitive or inductive coupling. If the wavelength and dimensions of the voltage carrying element are comparable, then radiative interference takes effect.

4 Interference Problems

4.1 Normal-Mode Interference

Normal-mode interference voltages appear along the outgoing and returning conductors of a circuit. They usually originate from either inductive coupling or from common-mode to normal-mode conversions (see section 4.3). Such disturbances overlap the operating signal.

Within balanced circuits, normal-mode interference voltages work as *symmetrical noise* (diagram 4.1), and consequently, as *unsymmetrical noise* in unbalanced circuits (diagram 4.2).

4.2 Common-Mode Interference

Common-mode interference is caused by interference voltages arising between the reference potential (e.g. signal ground) and all the conductors of a circuit. Within balanced circuits, common-mode interference voltages appear as unbalanced voltages between the electrical midpoint (arithmetical) of the circuit and its reference potential (signal ground). In unbalanced circuits, unbalanced voltages (asymmetrical noise or common-mode voltages) develop between the individual conductors and the reference potential (diagram 4.2). The unbalanced voltages (unsymmetrical noise) from the outgoing and returning conductors differentiate them-

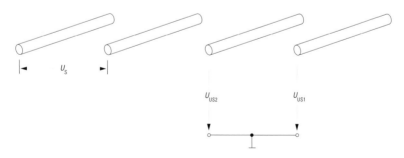

U_S Balanced interference voltage

U_{US1} Unbalanced interference voltage 1
U_{US2} Unbalanced interference voltage 2

Diagram 4.1 Balanced interference voltage (symmetrical noise)

Diagram 4.2 Unbalanced interference voltage (unsymmetrical noise)

selves in magnitude with respect to their effective voltages (differential-mode voltages).

The unbalanced voltage (asymmetrical noise) U_{AS} is the arithmetic mid-point between the unbalanced interference voltages (unsymmetrical noise) U_{US1} and U_{US2} influencing two conductors:

$$U_{AS} = \frac{U_{US1} + U_{US2}}{2} \tag{3}$$

The magnitude U_{AS} easily describes the interference voltages, provided that the unbalanced voltages (unsymmetrical noise) differ insignificantly and are large with respect to the balanced portion actually generating the interference voltages. Most reference literature does not differentiate between asymmetrical and unsymmetrical noise voltages and often refers only to the *common-mode* voltage.

4.3 Common-Mode to Normal-Mode Conversion

The explanations of common-mode interference may only be considered strictly valid for AC and DC circuits at low frequencies. With increasing frequency, the cable's inductance and especially its parasitic capacitance become very noticeable. As long as the respective impedances and parasitic capacitances are identical for the outgoing and returning conductors of a circuit, no interfering differential-mode (normal-mode) voltages result.

In the event of unequal impedances, the common-mode voltage drives currents of differing magnitude through the outgoing and returning conductors. Thereby, the voltages at the ends of the outgoing and returning conductors adopt unequal voltages with respect to the reference voltage. Thus, an interfering differential-mode (normal-mode) voltage may be generated by the original, non-interfering common-mode voltage. This conversion does not have to ensue entirely. The degree to which this transformation occurs is known as the *common-mode to normal-mode conversion factor CNC* (values range between 0 and 1).

The *common-mode to normal-mode attenuation CNA* is derived from the common-mode to normal-mode conversion factor.

$$CNA = 20 \log 1/CNC \tag{4}$$

Common-mode interference is often generated by unsymmetrical transmission in conjunction with ground loops through the common-mode to normal-mode conversion factor. The common-mode to normal-mode conversion factor via ground loops can be prevented by:

▷ increasing the impedance

▷ balancing the impedances of the outgoing and returning conductors

▷ protective shielding techniques and

▷ galvanically separating the transmission paths.

A balanced interference voltage arises between two conductors; whereas, an unbalanced interference voltage results between the individual conductors and the reference potential (diagram 4.1).

The following terms possess the same meaning as *normal-mode signal*:

Push-pull signal	Transverse voltage
Serial-mode signal	Balanced voltage
Differential-mode signal	Asynchronous voltage
Odd-mode signal	Differential voltage

The following terms possess the same meaning as *common-mode signal*:

Longitudinal voltage	Direct-axis voltage component
Parallel-mode signal	Unbalanced voltage
Even-mode signal	Synchronous voltage

Note:

The destructive limits of logic circuits are considerably lower for excessively balanced voltages than for unbalanced voltages.

4.4 Traveling Electrical Waves (Surges) in Power Supply Technology

An interesting effect occurring in power electronics should be mentioned here, since it can only be explained using wave theory. Wave theory is the basic principle for electrical transmission technology. Fast-switching insulated-gate bipolar transistors (IGBTs) are increasingly used in frequency converters for servo drives. They switch relatively large currents as quickly as possible. Typical switching times lie around the 300 ns range.

The switching process creates transient oscillations within the circuit. These oscillations possess a frequency spectrum reaching far into the MHz range. According to wave theory, backward and forward traveling waves take effect within the cable. The resulting voltages may, via their superimposition, reach double the amplitude of the original input switching voltage.

The insulating layer of the cables should be designed on the basis of these excessive voltages. For typical operational switching frequencies (e.g. 20 kHz), they must be considered as the continuous load.

4.5 Voltage Supply to Electronic Components

The lead wires supplying voltage to electronic components and modules should be sufficiently calculated for both static and dynamic loads.

During switching, peak currents may flow through the supply cables into the electronic components. These peaks may be several times larger than those present at the static state. Peak currents arise within the circuits themselves via recharging and shorting during the switching process. The conductor size of the lead wire should be chosen so that the voltage drop along the cable induced by these peak currents does not exceed the allowable circuit parameters.

Additionally, these voltage peaks occur with a frequency directly determined by the clock rate (operating speed) of the electronic components. Therefore, the supply line currents of modern fast-switching systems have a harmonic component, whose frequency may reach far into the MHz range. When choosing such supply cables, the aspects of EMC should absolutely be considered.

4.6 Clock-Signal Cables for Electronic Systems

In order to develop immunity against electromagnetic interference, synchronous technology is often employed within electronic systems. Thereby, the components stagger the switching of their input signals. An extra clocking signal triggers the switching process. This clocking signal is transmitted simultaneously to all components. The clock line must also be protected against the effects of electromagnetic influences. Otherwise, distorted clock signals may lead to erroneous behaviour of the component.

5 Electromagnetic Compatibility (EMC)

Electromagnetic compatibility expresses the ability of an electric or electronic device to function as expected within its electromagnetic environment without unduly influencing its surroundings.

This topic directly affects all designers and manufacturers of electrical equipment, systems and installations. First of all, the EMC ruling has as of 1. 1. 96 become law without restriction. Furthermore, the high processing and transmission frequencies of modern electronics demand exact consideration of their electromagnetic emissions and susceptibility. Noncompliance to the relevant EMC standards may be penalized.

Note:

By applying the EU conformity marking "CE" (diagram 5.1), the manufacturer declares that his labeled product conforms to *all* the EU-Guidelines which were applicable at the time of manufacture. Since the individual guidelines take effect at various dates, products of the same series produced in different lots may eventually have to satisfy different guidelines. Along with the EMC Guidelines, the "Low Voltage Guidelines" also exist.

Apart from these legal restraints, it should naturally be in the interest of every equipment or systems user to prevent electromagnetic influences from causing malfunctions or destruction.

The basic principle of EMC is that an electrical system has an effect on its environment via its electromagnetic emissions and conversely may simultaneously be affected by electromagnetic influences within its surroundings. EMC refers electromagnetic interference and susceptibility. Both may occur from conductive cabling or radiation (diagram 5.2).

Diagram 5.1 EU conformity declaration marking

The following terms and abbreviations are often present in technical literature relating to EMC topics:

EMI   Electromagnetic Interference   EMS   Electromagnetic Susceptibility
ESD   Electrostatic Discharge   EMP   Electromagnetic Pulse
LEMP Lightning Electromagnetic Pulse   NEMP Nuclear-Electromagnetic Pulse

Since the mechanisms of interference for emissions and susceptibility are nearly identical, it is sufficient for the scope of this EMC discussion to consider only one domain of effect. While it is true that relative results remain valid for both directions of interference, different boundary conditions must typically be considered for emissions and susceptibility. Therefor, the absolute values of results observed may, in fact, not be valid for both directions of interference.

Whether or not a disturbance is interfering depends entirely upon the EMC of the system in question. The identical disturbance may lead to unacceptable malfunction in one system whilst not impeding tolerable operation in another system. With regard to the opposing directions of transmission for electromagnetic interference, a system's EMC must always be seen within the context of a unified reference system: the interfering source, path, and sink. Eventual shielding or protection measures can, therefore, only be planned for a specific or similar reference system. Since the system's environment often exists as an incalculable interfering source, substitute environmental classifications are defined in the standards.

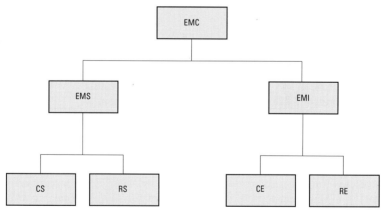

EMC   Electromagnetic Compatibility   EMI   Electromagnetic Interference
EMS   Electromagnetic Susceptibility   CE   Conducted Emission
CS   Conducted Susceptibility   RE   Radiated Emission
RS   Radiated Susceptibility

Diagram 5.2   Terms encountered with EMC

5.1 Environmental Classifications

The electronic systems must be capable of withstanding the electromagnetic influences present in the respective operational surroundings. The degree of noise immunity tests should correspond to these influences. For example, an *environmental class* has been assigned to the degree of the noise immunity tests stipulated for industrial control equipment in processing technology. According to the draft DIN VDE 0843 Part 4, the environmental class is determined through comparison with the actual installation and surrounding conditions. The environments are divided into four groups.

Environment class 1 (well protected environment)

The installation is characterized by the following attributes:
suppression of all pulse noise generated by switching the control circuits; separation of power supply, instrumentation and control cables connecting the system with an environment having a higher level of noise; shielded power supply cables with both end's of the shield connected to the installation's reference potential; as well as protecting the power supply with filters.

Computer rooms are an example of this environment class.

Environment class 2 (protected environment)

The installation is characterized by the following attributes:
partial suppression of pulse noise within control circuits; switching via relays (unprotected); separation of all circuits from those belonging to an environment with a higher level of interference; spatial separation of unshielded power supply, control, signal, and telecommunication cables; florescent lighting.

Control stations and terminal rooms in industrial installations or power stations are good examples of this environment class.

Environment class 3 (typical industrial environment)

The installation is characterized by the following attributes:

no suppression of pulse noise within switching circuits having relays (unprotected); minimal separation of switches belonging to an environment having a higher noise level; separate cables for power supply, control, signal, and telecommunication cables; availability of earthing systems via conduit cables, an earthing cable within cable canal (connected with the protective earthing system) and a general reference potential.

Installations for industrial processing, power stations, and relay rooms within outdoor, high-voltage sub-stations are examples of this environment class.

Environment class 4 (industrial environment with a high interference level)

The installation is characterized by the following attributes:
no suppression of pulse noise within the control and power supply circuits switched via relays and circuit-breakers; minimal separation of switches belonging to an environment having a higher noise level; no separation between power supply, control, signal, and telecommunication cables; common usage of multi-conductor cables for control and signal circuits.

Equipment for industrial processing external to industrial and power plants, where no unusual installation is foreseen. Outdoor and compressed-gas enclosed high-voltage switch gears up to an operating voltage of 500 kV (possessing its own installation guidelines) are examples of this environment class.

Environment class 5 (special)

A minor or greater electromagnetic separation between the source of interference and equipment, cables, switches, lead wires, etc. The quality of the installation may necessitate choosing a higher or lower environmental class. It is important that cables having a higher noise level may run through an environment with a lower noise level.

5.2 Reliable Transmission and Fault-finding

The electromagnetic effects acting upon the transmission system or cables possessing poor transmission characteristics can only generate one normal-mode or one common-mode interference voltage. Due to these two voltages, it is very difficult to determine the actual cause of interference, especially since most disturbances arise sporadically and are difficult to reproduce.

Careful planning of the entire system is the only aide in preventing interference from the start. Hereby, the applied transmission technology should satisfy the system requirements, and the EMC measures should be suitable for the system's environment. Unfortunately, the same design measures for cables often act in opposition when simultaneously satisfying these two goals.

5.3 EMC Law

With the law governing the electromagnetic compatibility of equipment, the legislator translated the "EMC Guidelines 89/336 EWG, dated May 3rd 1989" from the EU into German law. Along with these guidelines, this legislation contains: global descriptions, basic protection requirements, area of validity, and a system for evaluation conformity. However, no technical details are given.

The European Norms (EN) describe the technical details which must then be converted into national standards.

The key concept of the EMC law is user protection. The market shall only offer the user products, which are immune to third party electromagnetic influences, and which do not cause disturbances in other equipment by generating unallowable quantities of electromagnetic interference. On the one hand, limits for the interfering emissions are established so that the equipment still functions properly in the presence of permissible interference levels. On the other, the test parameters and methods with which the noise immunity of the equipment in question can be verified are defined. The norm is organized to accommodate both emission and immunity standards.

Components which cannot operate independently are not governed by the EMC law. This includes, for example, wires and cables. There exists no corresponding standard to define their EMC characteristics. Although, they may be an integral portion of an independently operating system with an adequately defined noise immunity.

Therefore, wires and cables for most electrical equipment should be definitely included in all EMC considerations. When contemplating noise emissions, few norms pay attention to the selection of wires and cables: for example, for industrial, scientific and medical high frequency equipment EN 55011 and DIN VDE 0875 Part 11 (noise suppression ISM), or for information processing equipment EN 55022 and DIN VDE 0875 Part 3 (noise suppression ITE). With respect to noise immunity, considerably more standards corresponding to the equipment to be tested exist.

In portraying this topic, it becomes evident that the legislator of the EMC law does not provide the manufacturer with instructions describing the design criterion for interference-free equipment and installations. This is recognizable by the large measurement ranges (e.g. 10 m to 30 m) and by the permissible limits defined for measuring the noise emission. Even when conforming to the limits, the field strength of the noise radiated in the immediate proximity of the emitting equipment are still relatively large. In this manner, for example, signal transmission over communication cables lying next to power cables may be impaired.

6 Cables for Signal and Data Transmission within Building Installations

When erecting transmission paths, the necessary cables should be evaluated by the following criterion:

▷ the switching technology integrated into the electronic system

▷ the suitability of the transmission medium

▷ the mechanical, thermal, chemical and electromagnetic robustness of the cable's design

▷ the technical suitability of the model and design of the cable's connector technology.

Principally, efficient transmission cables for conventional use within industrial and building installations demonstrate the following features or applications:

▷ balanced construction of twisted conductors in either pairs or quads

▷ conductor stranding with the shortest possible length of lay (approx. 10 to 30 mm)

▷ highly symmetrical cable construction (balanced cable) having tight tolerances and parameters used to define the symmetry

▷ highly homogenous material parameters and construction along the length of the cable. This implies a high longitudinal constancy of its transmission characteristics having exact median values.

▷ exclusive cable shielding employed only as protection against electromagnetic interference and not, for example, as a common return conductor for signal circuits

▷ transmission characteristics of the cable should be appropriate for the application requirements

▷ type and construction of the conductor contacts located at the cable's ends should also correspond to the demands of application

▷ simple and reliable means to terminate the cable's ends.

Further prerequisites for transmission of high frequency signals over long distances without interference include: e.g. establishment of the earth and reference potential, predetermination of the mechanical and thermal characteristics of the cable and terminations technology, etc.

6.1 Concentric Stranded Multi-Conductor Cables

Concentric stranded, multi-conductor cables are unbalanced cables (diagram 6.1). They are not suitable for symmetrical transmission. They are available in shielded or unshielded versions. Usually, parameters for the transmission characteristics are not given for concentric stranded cables. Typical cables for data and signal transmission available on the market have 2 to 100 conductors each with a diameter of approx. 0.4 to 1.8 mm or more. Concentric stranded control cables are described in the DIN VDE 0245 standard.

Concentric stranded cables have low EMC and are, therefore, only suitable for low transmission rates. Shields generally reduce, but do not eliminate, the susceptibility to electric fields. Shields for flexible cables are comprised of wire braiding having a degree of optical coverage always less than 100 %. The effectiveness of the shield usually improves with the degree of coverage. Common and economical values for coverage lie between 60 and 80 %. Thicker braided wires screen out magnetic fields better than thinner ones. The best shielding can be achieved with a solid ferromagnetic tube.

Static shields designated for fixed installation are comprised of plastic-coated aluminum foil and a parallel running drain wire. They offer 100 % coverage and a relatively high transfer impedance. The transfer impedance is an indicator of the shield's effectiveness against magnetic interference. The effectiveness of the drain wire decreases noticeably with the increasing frequency of the interfering field. For this reason, the drain wire should never alone be connected to the reference potential, rather the entire shield consisting of the foil plus drain wire (section 7.6). Further details concerning cable shields are discussed in chapter 7.

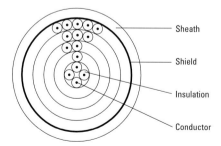

Sheath

Shield

Insulation

Conductor

Diagram 6.1   The principle construction of a concentric stranded, unbalanced cable

6.2 Balanced Cables

Balanced cables are most suitable for symmetrical transmission. Conventionally, copper wires are used. For high quality data cables, they must be bare (i.e. untinned). For balanced cables, a pair (side-circuit) is intended for each signal circuit.

Stranding elements for balanced cables may be made of pairs or quads. A *primary core unit* consists of several concentric stranded, stranded elements. Several primary core units are concentrically stranded to form a *major unit,* and several major units respectively form the *final core assembly.* This type of cable construction is called a *bunch stranded cable.* With *concentric stranded* cables, the stranding elements are organized in individual concentric layers. The *final core assembly* contains all the stranding elements of the entire cable, including the wrapping and its surrounding inner sheath.

Shielding may be found over the individual pairs and/or stranding elements and/or cores and/or final core assembly. Typical shields consist of braids and/or foils of the same types and attributes as previously described for concentric stranded multi-conductor cables. Diagram 6.2 depicts both the elements and the fundamental construction of communication cables.

Cables having bunch-stranded elements are easier to manufacture than concentric stranded cables and have comparable quality. This especially applies to cables having a large number of pairs. The individual units are not coupled to one another. Concentric stranded cables are generally thicker than comparable bunch-stranded versions.

The following Anglo-American designations are common for balanced cables including the non-standardized models:

TP	*Twisted Pair*	two insulted conductors stranded together
Q	*Quad*	four conductors stranded into a star-quad
STP	*Shielded Twisted Pair*	each individual pair is shielded
UTP	*Unshielded Twisted Pair*	the individual pairs are not shielded; however, an overall shield may be present

The letter F preceding or S *(screened)* proceeding indicates that an *overall shield* is present. It does not necessarily appear in the cable's designator.

Example

The designator UTP-S or FTP indicate a balanced cable consisting of twisted pairs and an overall shield.

Table 6.1   Typical cable parameters for transmission characteristics

German Parameter	English Parameter	Test Method
größte zulässige Frequenz	maximum specified frequency	–
größter Gleichstrom-Schleifenwiderstand	maximum DC loop resistance	IEC 189-1
größte Dämpfung	maximum attenuation	IEC 189-1
kleinste Ausbreitungsgeschwindigkeit	minimum phase velocity of propagation	
kleinste Nahnebensprechdämpfung (NEXT)	minimum near end crosstalk loss	DIN VDE 0472 Part 517
größter Widerstandsunterschied	maximum resistance unbalance	ffs
kleinste Erdunsymmetriedämpfung	minimum longitudinal to differential conversion loss	ITU-T O.121
Wellenwiderstand	characteristic impedance	IEC 1156
kleinste Rückflußdämpfung	minimum structural return loss	ffs
größte Erdkapazität	maximum capacitance unbalance pair to ground	IEC 708-1
größter Kopplungswiderstand	maximum transfer impedance	IEC 96-1

ffs: will be established in the norms at a later date

Along with the usual electrical and mechanical dimensions, parameters for the transmission characteristics of high quality cables need to be established. Table 6.1 shows a selection of the most important parameters necessary to check the suitability of a specific cable for a particular transmission requirement.

The parameters should either comply with the corresponding system requirements or be defined in the respective standards. The values are frequency dependent.

The following parameters are characteristic of good cable balance:

▷ unbalanced resistance

▷ near end crosstalk loss

▷ unbalanced capacitance with respect to ground

▷ longitudinal to differential conversion loss

▷ structural return loss.

The quality of the cable's balance primarily determines the EMC of the cable. It has a far greater influence than shielding. A symmetrical signal transmission with an ideally balanced cable (no inhomogeneity of the material compounds, negligibly small manufacturing tolerances) can not be disturbed by electromagnetic fields acting along the length of the cable. In this case a shield would be superfluous.

In reality, it is naturally not that simple. It is true that the values occasionally even suggest "factory fresh" cables without shielding; however, experience shows that the attribute of balance diminishes once the cable has been installed. Mechanical deformations result during installation. Since the shield acts as a bandage for the cable's core assembly. It protects the quality of balance and may improve the distribution of the other parameters along the cable. This effect is clearly observable in high quality cables designed for transmission in the 100 MHz range.

A profile of the requirements necessary for structured building cabling can be found in ISO/IEC DIS 11801 (Information Technology Generic Cabling for Customer Premises Cabling) as well as in DIN EN 50173. Balanced cables for communication and information processing systems are contained in the norm series DIN VDE 08... Furthermore, the norms for diverse electronic systems (e.g. bus systems) either state the specifications for the necessary cables or make reference to the appropriate cable standard.

The permissible length is directly determined by the transmission frequency in conjunction with transmission characteristics of the cable. Thus, the maximum permissible cable lengths described in ISO/IEC DIS 11801 are also stated as a function of transmission frequency (see section 2.6).

The merits of balanced cables during symmetrical operation compared with unbalanced cables are:

▷ a far smaller common-mode/normal-mode conversion factor

▷ lower sensitivity to magnetic interference fields

▷ no galvanic coupling between the individual circuits

▷ a shielded cable is only necessary for high transmission frequencies

▷ no possible galvanic coupling via ground loops

▷ better transmission characteristics for cables having a large number of conductors

▷ minimal electromagnetic emissions.

These advantages lead to a noticeable improvement in EMC for symmetrical transmission in comparison to unsymmetrical transmission.

A cost comparison between flexible, unbalanced control cables with a small number of conductors and flexible, balanced control cables having a comparable number of conductors is listed in Table 6.2.

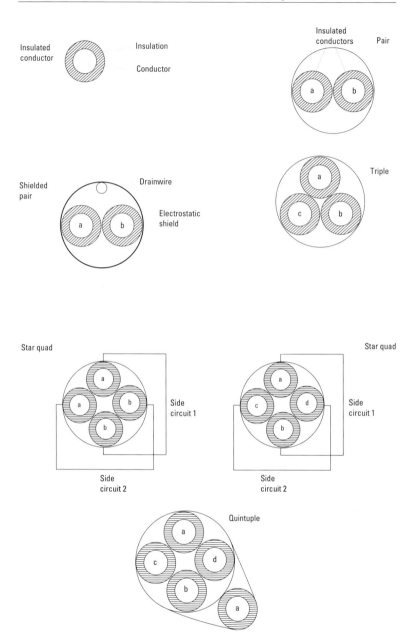

Diagram 6.2 Basic construction and elements of communication cables

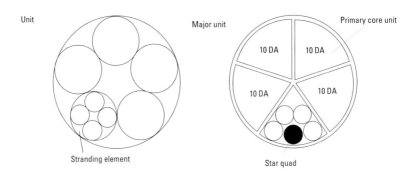

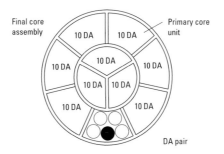

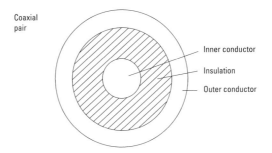

Table 6.2 Cost comparison for cables having two side-circuits

	Cost Factor
Multi-conductor control cable $3 \times 0.25\,\text{mm}^2$ (unshielded)	1
Multi-conductor control cable $3 \times 0.25\,\text{mm}^2$ (as above, however with braided shield)	2.4
Control cable $2 \times 2 \times 0.25\,\text{mm}^2$ (as above, however with twisted pairs, unshielded)	1.8
Control cable $2 \times 2 \times 0.25\,\text{mm}^2$ (as above, however with twisted pairs and braided shield)	2.9

As long as the cables do not transmit at the highest of frequencies, the quality of shielded, unbalanced cables is comparable to that of unshielded, balanced cables. Therefore, a shielded, unbalanced cable may be more expensive than an unshielded, balanced cable of similar quality. The additional expenditures for installing the cable's shield is not even included in this consideration.

For the highest of frequencies, only balanced cables are suitable. The minimum attributes of high quality balanced cables include:

▷ the shortest length of lay for stranded pairs or quads (approx. 10 ... 30 mm)

▷ mechanically stable construction and/or expensive shielding

▷ PE- or cell PE Insulation

▷ mechanical and electrical compatibility with the anticipated installation system (e.g. fastening technologies).

In order to evaluate and select the appropriate transmission cables, planners and users need at a minimum the following transmission information from the suppliers:

▷ the standard governing the cable or

▷ detailed information concerning the cable's parameters.

Since most system standards provide only the cable's profile, it is possible that lower quality cables may satisfy this profile, but be noticeably different in other non-specified attributes. Primarily cost and performance comparisons must be considered in this circumstance (see section 2.6). The appendix contains a check list which enables not only a comparison but also a fixed determination of the relevant and required dimensions and characteristics for signal cables.

Within building technology, signal and data transmission cables are largely needed for systems and installations.

Systems

Structured cabling systems
Building control engineering
Air-conditioning and regulation
Security systems

Building system technologies
Building management systems
Telecommunications
Local networks for computers and
similar systems

Installations

Industrial installations
Peripheral processing
equipment
Local networks for computer
integrated manufacturing CIM

Machines
Instrumentation and control
technology

These applications predominantly require bus cables or cables for token ring networks.

6.3 Coaxial Cables

The transmission characteristics for balanced cables have been drastically improved. The cables are now suitable at transmission frequencies extending into the 100 MHz range and higher over the common transmission distances found within buildings. From the present vantage point, little future demand can be seen for expensive, unbalanced coaxial cables outside of high definition television (HDTV). They are, therefore, not a conventional component for building systems. Even though coaxial cables are still partially stipulated for some LAN applications, it is usually out of historical reasons. At the time these requirements were determined, coaxial cables were the only type available exhibiting the necessary performance features. Besides, in the event that balanced cables no longer suffice, the use of fiber optics is less expensive and offers more advantageous EMC.

6.4 Fiber Optic Cables

Within buildings, the use of mineral glass fiber optic cables only makes sense for the risers (vertical cabling) or for extreme EMC demands. Furthermore, polymer optical fibers are recommended for applications in which the transmission advantages and improved EMC justify the additional financial expense.

The performance capabilities of balanced copper cables is far greater than those presently required for building installations. Therefore, the use of polymer optical fibers is temporarily restricted to special applications.

The standards for future-oriented cabling structures for building installations also view fiber optics as an alternative to balanced copper cables only for horizontal applications.

6.5 Cable Parameters for Transmission Characteristics

Cable parameters describe the mechanical, electrical and transmission characteristics. All parameters for transmission characteristics are frequency dependent and often have complex dimensions.

Characteristic impedance

The cable manufacturer generally only gives the value (Ω) of the complex characteristic impedance. It is independent of the cable's length. For an ideal cable, it is identical at all locations. The imaginary portion approaches zero with increasing frequency (above approx. 1 MHz). High frequency cables are operated at frequencies having a real characteristic impedance. This value is often used as an indication of the cable type: e.g. 100-Ω-cable. Typical values for multi-conductor, balanced cables are 100 Ω, 120 Ω and 150 Ω.

If a cable is not terminated with a resistance equivalent to its characteristic impedance, incident waves will be reflected at this resistance. Inhomogeneities along the cable locally change the value of the characteristic impedance, which likewise lead to reflection. The characteristic impedance of a good cable not only has tight tolerance boundaries but also demonstrates minimal deviations along the cable's length. A measure of the continuity, is the structural return loss. This value should be as large as possible.

Loop resistance

The loop resistance (Ω) is the sum of the resistances for the outgoing and returning conductors of a circuit. For balanced cables, these are the conductors of a pair (side-circuit). The loop resistance usually refers to a particular length of cable (e.g. 1000 m).

Unbalanced resistance

The unbalanced resistance is the difference (%) between the ohmic resistances of the outgoing and returning conductors (e.g. the conductors of a loop). It should be kept small.

Wave attenuation

The wave attenuation α is the frequency dependent reduction (dB) in the amplitude of a wave after traveling through a conductor section. It usually refers to a particular length of cable (e.g. 100 m). The wave attenuation should be as small as possible.

Phase velocity

The phase velocity is the speed at which a sinusoidal wave with a particular frequency propagates along the cable. It is given in km/s or as a portion of the frequency of light (relative phase velocity of propagation). High quality cables have a higher phase velocity.

Mutual capacitance

The mutual capacitance is determined by the capacity between the conductors of a pair (loop) and the capacitance between the conductor and the shield of a balanced cable. The length of the cable and the measurement frequency is defined. It is stated as a function of length (e.g. nF/km). The mutual capacitance should be kept as small as possible.

Transfer impedance

The transfer impedance is a good measure of the effectiveness of a cable's shield against magnetic influences. It is therefore important for EMC considerations. It refers to a particular length of cable (e.g. mΩ/m). The transfer impedance should be kept as small as possible and constant over the frequency range.

Near end crosstalk loss

The near end crosstalk loss (*NEXT*) is a frequency dependent measure (dB) for estimating the potential for crosstalk. It refers to a particular length of cable (e.g. 100 m). The value should be as large as possible.

Attenuation-to-crosstalk-loss-ratio

The wave attenuation continually diminishes the amplitude of a signal along the cable. Simultaneously, interfering signals from neighboring cables are coupled into the cable via crosstalk. As long as the separation (as a function of the transmission frequency) between the wave attenuation and the near end crosstalk loss is large enough, reliable transmission can be expected. If it is too small or even negative, then the transmission is unreliable. The attenuation-to-crosstalk-loss-ratio (*ACR*) is determined from the difference in the wave attenuation α and the near end crosstalk loss (*NEXT*):

$$ACR = NEXT - \alpha \quad \text{in dB} \tag{5}$$

The *ACR* is frequency dependent and becomes smaller with increasing frequency. In order to achieve bit error rates of less than 10^{-12}, a typical *ACR* of min. 12 dB is necessary.

Minimum longitudinal to differential conversion loss

Due to the unbalanced capacitances to earth for the outgoing and returning signal conductors, the transformation of common-mode voltages into normal-mode voltages partially occurs. Usually, the less interfering common-mode voltages are coupled into the circuit. Their transformation into interfering differential voltages is highly noticeable during signal transmission. The minimal longitudinal conversion loss (*LCL*) is given in dB. This parameter should remain constant over the entire frequency range.

Structural return loss (reflection attenuation)

Electric waves may be partially reflected in varying degrees at discrete locations along the cable wherever discontinuities in the wave attenuation occur. The reflected waves overlap with the operating signal. A measure for the homogeneity of the characteristic impedance along the cable is the structural return loss α_R. It can be determined from the characteristic impedance Z_x and the terminating impedance Z_0 of the cable:

$$\alpha_R = -20 \log \left| \frac{Z_K - Z_0}{Z_K - Z_0} \right| \text{ in dB/100 m} \tag{6}$$

If the rated value of the characteristic impedance is substituted for Z_0, the "return loss" can be determined. If the average measured characteristic impedance is applied, then "structural return loss" can be determined. The structural return loss should be as large as possible.

All parameters for the transmission characteristics are clearly responsive to frequency and temperature.

ISO/IEC DIS 11801 is an important norm for building installation technology. Herein, specifications and types of cables having the classification "cable category 3 ... 5" are given. Table 6.3 contains an excerpt.

As the values in the table distinctly show, the parameters vary throughout the cable category not only at the frequency limits, but over the entire frequency range.

Table 6.3

Excerpt for parameters of the transmission characteristics for 100-Ω cables according to ISO/IEC DIS 11801

Parameter	Dimension	Frequency MHz	Cable Category 3	Cable Category 4	Cable Category 5
Characteristic Impedance Z	Ω	0.064 ≥ 1	$125 + 20\%$ ffs	$125 + 20\%$ ffs	$125 + 20\%$ ffs
Maximum Loop Resistance	Ω	DC	19.2	19.2	19.2
Maximum Unbalanced Resistance	%	DC	3	3	3
Minimal Near End Crosstalk Loss *NEXT* (cable length 100 m)	dB	0.772 4 10 16 100	43 32 26 23 –	58 56 41 38 –	64 62 47 44 32
Maximum Wave Attenuation α	dB/ 100 m	0.064 0.772 1 4 10 16 100	0.9 2.2 2.6 5.6 9.8 13.1 –	0.8 1.9 2.1 4.3 7.2 8.9 –	0.8 1.8 2.1 4.3 6.6 8.2 22
Minimal Structural Return Loss α_R (cable length 100 m)	dB	$1 \dots < 10$ $10 \dots < 16$ $16 \dots < 20$ $20 \dots < 100$	12 10 – –	21 19 18 –	23 23 23 $23 - 10 \log (f/20)$

DC direct current f frequency   ffs under discussion

7 Shields for Cables

The designer can influence a variety of cable attributes by using various components for the shields (also see section 6.1). Possibilities include:

▷ transmission characteristics of the cable

▷ torsional strength

▷ resistance to repeated bending (flexural strength)

▷ the cable's overall flexibility.

If the shield components provide protection against external mechanical stress, they are called armouring. The following discussions deal solely with a shield's effectiveness against electromagnetic interference (both emissions *and* susceptibility).

The German telecommunications board requires, for example, in FTZ TL 6145–3000 shielding for all cables operated at a frequency above ≥ 1 MHz. Principally, the effectiveness of a shield can only be quantitatively judged in connection with its respective transmission system:

▷ transmission mode (symmetrical or unsymmetrical)

▷ cable type (coaxial, twisted pair, stranded insulated conductors)

▷ cable parameters (length of lay, near end crosstalk loss and characteristic impedance)

▷ signal-to-noise ratio of the interconnected electronic circuits

▷ transmission and interference frequencies

▷ single-ended vs. double-ended earthing of the shield

▷ multi-functionality of the shield

▷ terminating resistance of the cable, etc.

Beyond this, the effectiveness of the shield against the individual electromagnetic influences usually varies and is always frequency dependent. Since transmission frequencies for building systems technology have meanwhile reached the 100 MHz range, the shields employed for data transmission cables must be examined for the aspect of high frequency operation.

Shields always alter the original transmission characteristics of the cable. For example, shielded cables have a higher mutual capacitance than similar non-shielded versions.

Completely seamless, closed shields are more effective than porous shields (e.g. plastic-coated aluminum foil shields). The better the conductivity of the shield construction and its entire circuit, the larger its shield current and, thereby, its effectiveness. For this the reason, copper is a better shielding material than steel at frequency ranges under 100 kHz.

Magnetic fields at frequency ranges under 100 kHz penetrate the walls of conventional cable shield designs. Therefor, only the field generated by a current can help to counteract the interfering field and can to a large extent compensate for it (section 7.2.2). This current itself is induced within the short-circuited shield by the interfering field through double-ended earthing. For frequencies above 100 kHz (e.g. electromagnetic waves), the penetration depth of the fields is often less than the thickness of the customary shield (section 7.2.4). The penetrated waves become absorbed. Steel is a better shielding material for this purpose than copper.

Inhomogeneities, openings or leaks in the metal shields act as aerial frames (a transmitter of magnetic fields) within a protected surrounding.

The following is particularly important for shields:

1. A shield that is not earthed on both ends has minimal or no effectiveness at low frequencies (under the MHz range).

2. At high frequencies the shield itself acts as a transmitting/receiving antenna.

In all applications, a sealed ferromagnetic tube, in which the insulated conductors of a circuit are placed, provides the best shielding results. Unfortunately, such shielding is extremely expensive and not practical for most applications. Suitable components should be incorporated into the cable's design.

7.1 Components for Cable Shields

All materials exhibiting a high conductivity for the particular electromagnetic field or having the capability of producing a compensating field are suitable for shielding purposes.

According to DIN 57 271 A3, metal bands or wires surrounding the individual or stranded conductors are viable shields. If a copper braided shield is present, then it is designated in the telecommunications nomenclature by the symbol "C". Shielding from metal bands or plastic-coated metal bands or foils have the symbol "St" (static shield). Plastic-coated metal bands generally surround the element in the form of an overlapping spiral. The individual windings of the spiral are short-circuited (connected via minimal resistance) with a bare drain wire wrapped along with the bands. Shields having longitudinally running foils also exist. Since the edges of the foil are joined with a folded seam, it creates a quasi-sealed metal tube. A drain wire is not necessary.

Static shields are generally found in cables destined for fixed installation. Braided shields, though more expensive, are better suited for flexible applications. For highly flexible applications, only cables having spun wires are suitable. Unallowable, frequently repeated bending of the cables leads to lasting deterioration of the shielding effectiveness. The best shielding is achieved, when the shields are only employed for their electromagnetic shielding function, instead of additionally acting as a current carrying conductor.

7.2 The Functionality of a Shield

Shields increase the mutual capacitance. They act similarly against electromagnetic emissions and susceptibility. Their varying functions with the different coupling methods are described in the following sections.

7.2.1 Protection against Capacitive Coupling

A shield protects against electric fields when it possesses the following attributes:

▷ a negligibly small coupling impedance

▷ a negligibly small transfer capacitance

▷ an ideal conductivity

▷ no inductivity.

The field lines emanating from the interfering system end at the shield. The dielectric current flows over the earthed shield to the interfering source and back, instead of flowing along the conductor and through the connected electronic system (diagram 7.1). Thus, no interfering voltages occur in this system.

As braided shields are especially "leaky", a transfer capacitance must be expected. Currents flow through the shield into the cable's interior via this capacitance. There they are carried along the conductor into the electronic system and cause interference voltages.

A non-earthed shield or high shield impedance leads to an increase in the shield's potential, and due to stray capacitances, it also leads to an interfering current flow via the conductor to the connected circuits. A single-ended earthed shield only protects against low frequency electric fields having a wave length larger than seven times the cable's length. The shields effectiveness against capacitive influences decrease starkly as the frequency increases.

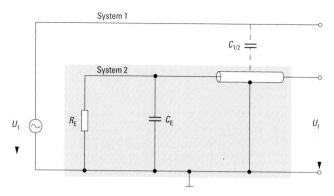

U_1 Voltage in system 1
U_I Interference voltage in system 2
R_E Replacement resistance for the internal resistance of the transmitter and receiver in system 2
C_E Replacement capacitance for the capacitance of transmitter and receiver of system 2
$C_{1/2}$ Parasitic capacitance (stray capacitance) between system 1 and system 2

Diagram 7.1 Functional principle of a cable's shield against capacitive coupling

7.2.2 Protection against Inductive Coupling

As previously mentioned, magnetic fields penetrate the typical thickness of a shield. Magnetic interference fields are combated by building an opposing magnetic field (diagram 7.2).

Double-ended earthing of the cable's shield develops a closed circuit. An external, alternating, magnetic field induces an opposing electromotive force (EMF) within this circuit. Its generated current produces a magnetic field in the opposite direction which weakens or partially compensates the interfering field. In order to achieve total compensation, large currents must flow. This requires a low shield impedance. The smaller the impedance of the entire shielding circuit is, the better the protection provided against inductive coupling. Shields made of foils, thin wire braiding or open braids usually do not offer sufficient shield impedance.

A single-ended earthed or non-earthed shield does not protect against inductive coupling, because an open circuit does not allow compensation current to flow. At high frequencies, the stray capacitances to earth are large enough to form a closed circuit. Then a shield current can flow and create a magnetic field to oppose the interference field.

If the galvanic earthing of the second shield side is not permitted (section 7.2.3), a connection can be established by using a compensation capacitor.

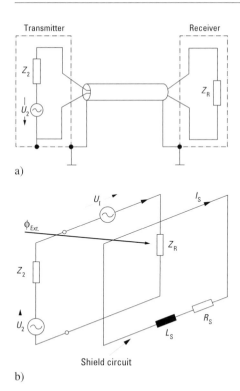

Transmitter   Receiver

a)

b)

Shield circuit

U_2 Voltage in system 2
U_I Interference voltage in system 2
I_S Interfering current in the shield circuit
Φ_{Ext} External alternating, interference field
L_S Inductivity of the cable's shield
R_S Resistance of the cable's shield
Z_2 Impedance of the transmitter in system 2
Z_R Impedance of the receiver in system 2

Diagram 7.2   Functional principle of a cable's shield against inductive coupling
a) Layout   b) Operational schematic

7.2.3 Protection against Galvanic Coupling

Cable shields do not protect against galvanic coupling. Galvanic coupling arises when the shield is employed as a common return conductor (reference potential) for several circuits.

When the potentials of a double-sided earthed shield differ, large, high frequency compensation currents may flow along the shield. Voltage drops arise in the form of normal-mode interference due to the shield's transfer impedance (section 4.1). This type of ground loop should be undone at a single point. However, the cable's stray capacitances to earth still remain and an electronic activation exists. At high frequencies, they have relatively small impedances and act as galvanic connections to earth. Thus, ground loops remain at high frequencies.

7.2.4 Protection against Radiative Coupling

As both magnetic and electric components with high frequencies are present, only difficult calculations using antenna theory are possible. In comparison to capacitive coupling, one distinguishes between two distinct electric fields. One stands perpendicular to the cable's shield, and the other runs parallel to the cable's axis. The incident wave creates a high frequency current in the walls of the conductive shield. These in turn generate magnetic fields. Due to the high frequency, the shield itself radiates electromagnetic waves.

The function of a metallic shield against electromagnetic radiation can be explained as follows:

When an electromagnetic wave is transferred from one medium into another having a different characteristic impedance, partial reflections occur at the junction. An electromagnetic wave experiences the same effect, when it comes in contact with the shield. The shield reflects a portion of its energy. The remainder penetrates the shield and is attenuated. If the shield is thick enough, then the remaining energy will be absorbed. No radiation remains to escape through to the interior of the cable. Radiation richer in energy does reach the inside wall of the shield. Here, it is again partially reflected upon contact, and a small remainder passes through to the interior. The effectiveness of a cable's shield is, therefore, founded upon absorption and reflection principles. "Leaks" in the shield, such as those always present in braided versions, reduce the shielding ability and allow the waves easy penetration.

The penetration power of an electric wave decreases with increasing frequency due to the skin effect. The measure of penetration (table 7.1) gives the penetration depth with which an electromagnetic wave is weakened to a portion (e^x) of its original energy (before it enters the shielding material).

For practical applications, a shield thickness of approximately five times the penetration depth δ is sufficient to internally absorb the interfering waves.

A further effect of radiative coupling should be examined. At very high frequencies, shield sleeves behave like resonant cavities (resonance chambers). At frequencies approaching the intrinsic resonance, resonance notches occur in the shield's attenuation. The attenuation is drastically

Table 7.1 Penetration depth δ for various materials and frequencies

Frequency	Penetration depth δ		
Hz	Copper mm	Aluminum mm	Iron mm
50	9.44	12.3	1.8
10^2	6.67	8.7	1.3
10^3	2.11	2.75	0.41
10^4	0.68	0.87	0.13
10^5	0.21	0.28	0.041
10^6	0.068	0.087	0.013
10^7	0.021	0.028	0.0041
10^8	0.0068	0.0087	0.0013
10^9	0.0021	0.0028	0.0004

reduced to such an extent that the electromagnetic interference fields pass relatively unsuppressed through the shield walls. This is known as resonance catastrophe (section 7.5). Otherwise, the same measures may be implemented to reduce the radiative coupling, as those used against electric and magnetic coupling (e. g. shields, stranding, etc.).

Table 7.2 depicts the effective tendency of cable shields against the various types of coupling.

Table 7.2 The effectiveness of various cable shields having an optimal construction

	Low Interference Frequencies	High Interference Frequencies
Inductive Coupling – single-sided earthing – double-sided earthing	– ++	+ ++
Capacitive Coupling – single-sided earthing – double-sided earthing	+ ++	– ++
Galvanic Coupling – single-sided earthing – double-sided earthing	0 0	0 0
Electromagnetic Coupling – single-sided earthing – double-sided earthing		+ ++ Possible resonance catastrophes

++ very good shield effectiveness
+   good shield effectiveness
–   less effective shielding
0   no shielding effectiveness

Table 7.3 Capacitance and resistance values for capacitive shielding

Cable length	C_M and C	R
m	µF	Ω
5	100	0.22
20	1	1
50	0.1	4.7
100	0.01	22
200	0.022	100
500	0.047	100
1000	0.1	100
2000	0.22	100
5000	0.47	100

7.2.5 Capacitive Shielding

In cases such as those described in section 7.2.3, where double-sided earthed shield may not be employed, capacitive shielding may be substituted as useful solution for most applications. It is necessary to galvanically connect one end of the shield directly and the other side via a capacitor C_M to earth (diagram 7.3). In order to avoid resonances it is advisable to connect the capacitance C_M in parallel to an RC circuit element. Thereby ensuring that the capacitor's value is as large as that of C_M. Depending on the length of the cable, the values for R, C and C_M should be chosen according to table 7.3.

The galvanic connection to earth should be made at the appropriate end of the shield according to the following:

▷ for potentially connected transmission at the receiver end

▷ for potentially separated transmission at the potentially connected end

▷ for symmetrical transmission at the end, where the conductor has a lower impedance with respect to the reference potential

▷ for transmission over coaxial or triaxial cable systems at the end, where the outer conductor of the coaxial system has a lower impedance with respect to the reference potential

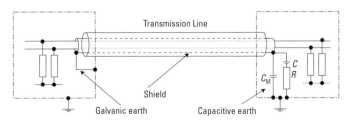

Diagram 7.3 Capacitive shielding

Capacitive shielding achieves good shield effectiveness against higher frequency electromagnetic interference. However, the capacitance C_M reduces the effectiveness against low frequency interference. Capacitive shielding is generally preferable to a single-sided earthed shield.

7.3 Types of Shields

Braided shield

Braided shields generally consist of copper wires. By painting a thin silver coating over these wires, the skin effect at the surface is minimized. Flexible cables require braided shields. Principally, the denser the braiding, the more effective the shield, but the stiffer the cable. A braided shield which is totally impervious to electromagnetic influences can not in practice be realized.

Foil shield

Foil shields are available in several varieties. The active shielding material is usually aluminum and occasionally copper. A difference between bare foil and plastic-coated foil exists. The former consist of a homogeneous metal layer, and the latter of a plastic foil base which is coated either single- or double-sidedly with a very thin metal layer only a few μm thick. The foils are applied in bands. One possibility being a spiral, overlapping wrap. The other is a longitudinal casing running along the cable's length. For plastic-coated foils, the edges are folded so that the metallic sides are in contact forming a closed, galvanic shielding sleeve. The edges of the foil for a spiral wrapping cannot be galvanically sealed.

Within the cable, a metallic drain wire runs along the metallic side of the foil to directly shunt any interfering currents. If it must be incorporated underneath the shield, then it becomes ineffective at higher frequencies. Due to the skin effect, the interfering currents can only flow along the surface of the shield above. The skin effect becomes apparent at frequencies of several 10 kHz. At high frequencies, the transfer impedance of foil shields is many times to the tenth degree higher than that of braided shields. However, foil shields are not suitable for flexible cables. The effectiveness against magnetic interference fields can be greatly improved by including ferromagnetic materials.

Foil shields may not be glued to the inside of the cable's outer sheath (e.g. for the purpose of an earth contact). Otherwise, they may rip whilst stripping the outer jacket. Such cable's are useless for making shield connections along its entire length (section 7.6).

Alongside the common braids and foils, other types of shield exist for demanding applications.

Spun shielding

Spun shield gives a cable high flexibility, but thereby reduces its shielding effectiveness. Spun shields consist of fine wire, which lie parallel to one another and are wrapped together around the core assembly in a spiral.

Double shielded cable

A double shielded cable has two similar concentric shields galvanically separated and arranged over each other. The outer shield of a coaxial cable serves as mechanical protection, and the inner as a signal return conductor. Hereby galvanic coupling with the shield is avoided. The outer shield is always earthed at both ends. A single-sided earthing of the inner shield yields better effectiveness at low frequencies. Consequently, double-sided earthing produces better effectiveness at higher frequencies. However, the latter leads to ground loops within the circuit and is not often recommended.

Multi-layered shield

Multi-layered shields are made of a variety of shield types concentrically arranged and galvanically connected. With this approach, the advantages of one shield type compensate for the disadvantages of another over the entire frequency range. Popular combinations include a braided shield over a foil shield or a braided shield sandwiched between two foil shields.

Shields should be viewed as a cable component. The closer the construction of a shield approaches that of a sealed metal tube, the better its characteristics.

7.4 Parameters of Cable Shields

All values and parameters of a cable's shield are frequency dependent. As previously discussed, their effects on the noise immunity of a transmission path may only be assessed in conjunction with all attributes (e. g. cable termination) of the transmission path.

Transfer impedance

The transfer impedance is the relationship between the galvanic or inductively coupled interference voltage on the inside of a cable's shield and interfering current introduced on the outside of the shield. The transfer impedance should be as small as possible (several $m\Omega$). For coaxial cables, the transfer impedance is an indicator for the susceptibility to magnetic and galvanic coupling. It yields little information for multi-con-

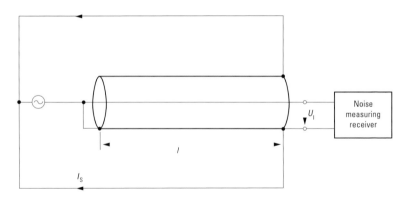

l   Cable length
U_I Interference voltage on the inside of the shield
I_S Interference current within the shield

Diagram 7.4   Test construction used to determine the transfer impedance

ductor cables. Diagram 7.4 depicts the test construction used to determine the transfer impedance.

Diagram 7.5 demonstrates the frequency dependency of the transfer impedance for several shield types.

Similar plots may only be practically expected, if the shield is optimally terminated (section 7.6). Inadequate terminations erode or even completely eradicate the shield's effectiveness especially at higher frequencies.

Shield factor

The shield factor is the relationship between an external interference voltage effective on the shield and the voltage between the inside of the shield and an internal conductor. The test construction is identical to that of the transfer impedance. The shield factor should be as large as possible.

Derating factor

The derating factor is the reciprocal of the shield factor.

Shield attenuation constant

The shield's attenuation constant is twenty times the logarithmic value of the shield factor.

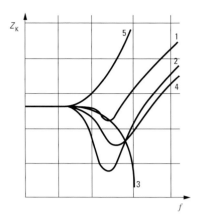

1 Single braid	Z_K Transfer impedance
2 Double braid	f Frequency
3 Sealed tube	
4 Tube having a 0.2 mm wide lengthwise gap	
5 Spiral overlapping bands	

Diagram 7.5 Plots of the transfer impedance for various shield types

Transfer admittance

Whereas the transfer impedance refers to inductively coupled interference, the transfer admittance is a measure of the shield's susceptibility to capacitive coupling. It is not determined by the shield alone, rather from the interfering environment. Diagram 7.6 shows which variables are used to determine the transfer admittance. The capacitance $C_{1/2}$ represents the disturbing environment.

The transfer admittance is also the relationship between the interfering current flowing along the shield and the induced interference voltage. Since the intensity of the magnetic field encountered in practice is generally larger than that of the electric field, the transfer admittance has been hitherto less important.

Transfer capacitance

The transfer capacitance k is not frequency dependent and can be determined from the length of the shield L and the capacitances $C_{1/2}$, C_1 and C_2 (diagram 7.6). Also see DIN 47 250 Part 4.

$$k = C_{1/2} \cdot L / (C_1 \cdot C_K) \tag{7}$$

79

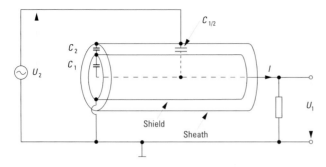

U_1 Resultant voltage
U_2 Test voltage
C_1 Partial capacitance between the conductor and shield
C_2 Partial capacitance between the shield and environment
$C_{1/2}$ Transfer capacitance of shield

Diagram 7.6 Parameters used to determine the transfer admittance

Shielding Unit

The shielding unit is a measure of the shield's ability to attenuate electromagnetic waves. It is defined as ten times the logarithmic relationship between the power fed into a shield and its outwardly radiated portion (also reference DIN 47 250 Part 6).

7.5 Resonance Catastrophe

When an external electromagnetic wave strikes a non-ferromagnetic metal shielding tube, the magnetic field component stands perpendicular to the shield's surface, and the electric component runs parallel to the tube's axis. At particular frequencies, resonance catastrophes occur. For the outer radius of the cylindrical shield r_a and the wavelength λ, magnetic field components first lead to resonance catastrophes when $r_a/\lambda = 0.6$. Further resonance points are reached at intervals of $r_a/\lambda = 0.5$.

Electric field components cause the initial resonance catastrophe at approx. $r_a/\lambda = 0.4$. Further points also occur at intervals of $r_a/\lambda = 0.5$.

Example

For a cable with a diameter over the shield of $2r_a = 6$ mm, the first resonance catastrophe via magnetic field components appears at approx. 60 GHz. Further resonance catastrophes follow in intervals of 50 GHz at 110, 160, 220 GHz and so forth. The electric field components induce the first resonance catastrophe at 40 GHz.

During resonance catastrophe, the shield's attenuation breaks down and a large portion of the external field components penetrate the cylindrical shield tube. If the waves incident the metal shield tube from other directions, the effects of the magnetic field components do not change. The electric components change only slightly.

Mobile communication networks operate at frequencies between 0.146 and 1.8 GHz (see appendix). For entirely, galvanically sealed cable shields with the above mentioned interior diameter, resonance catastrophes do not pose a threat. Sealed cable shields having seams behave differently. Due to their construction, resonance catastrophes are possible at the above mentioned frequencies. Further research into this topic is not known; however, recent practical experience indicates such a potential exists.

7.6 Techniques for Connecting Cable Shields

An important attribute for cable shields is the type and quality of the cable's termination. It's impedance should be as small as possible to enable large currents to flow along the shield. The common rule applies: the higher the frequency, the greater the importance of and expenditures for terminations should be.

The shields should generally be terminated across large surface area and around the entire circumference of the cable. Twisting the ends of the shield into plait, increases the shield's impedance to such an extent, that it hardly remains effective against inductive coupling. Static shields (e.g. thin plastic-coated foils) must be prepared so that a large surface area of the shield around its entire circumference can be terminated together with the drain wire. Simply terminating the drain wire is virtually useless. Additionally, the path of connection from the shield to the earthing potential should be as short as possible (a few centimeters). The twisted pairs should remain stranded along the unshielded portion of the cable's path.

Diagram 7.7 portrays a good and a poor example of arranging braided shield terminations.

Diagram 7.8 a and diagram 7.8 b depict the arrangement of the foil shield ends for termination. In this sketch, the foil shield is re-wrapped back over the outer sheath with the metallized side facing outward. For the preparations preceding the end of the cable, the foil shield is first cut open. Then the drain wire is pulled out of the core assembly to form a loop which is then wrapped around the core.

The shield ends must, of course, always be properly handled during insertion into sockets and connectors, as the effectiveness of a shield along a transmission path is determined by its weakest link.

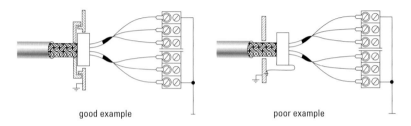

good example   poor example

Diagram 7.7   Examples for the preparation of the shield ends for termination

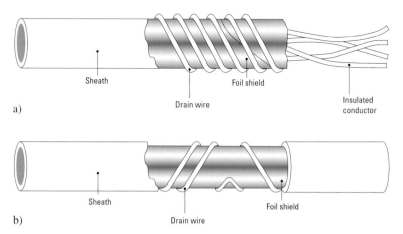

a)

Sheath

Drain wire

Foil shield

Insulated
conductor

b)

Sheath

Drain wire

Foil shield

Diagram 7.8   Examples for the preparation of foil shields for termination

7.7 Shield Terminations within Metal Housings

The cable's shield should be earthed at or in the immediate proximity of the cable's inlet through the wall. Otherwise, the shield could act as a noise transmitter within the metal housing (e.g. switching cabinet). This is advisable for all shields contained in a bus bar (e.g. flat copper bar, fanning bar or serrated track bar). These bars should be earthed over a large surface area. They should not be employed as cable clamps. If the cable must be extended over a short length without interruption within the housing, then the shield should only be earthed once, at the inlet through the wall, and not again at the cable's end. For this purpose, the foil shield should be connected around the entire circumference of the cable at the inlet. Merely attaching the drain wire is insufficient. If the non-metallic side faces outward, the drain wire should be drawn far enough out of the unsheathed portion of the cable so that it may be wrapped in a

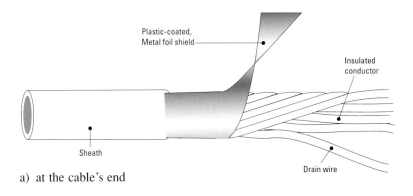

a) at the cable's end

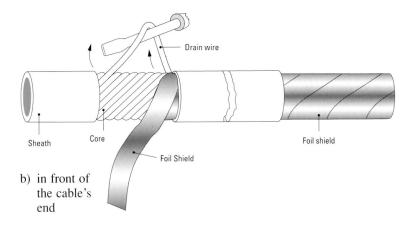

b) in front of
the cable's
end

Diagram 7.9 Preparation of a plastic-coated metal foil for contact around the entire circumference of the cable

loop around the core assembly. This "winding braid" is suitable for an all-around contact (diagram 7.9). For longer paths, it might be advantageous to earth the shield a second time at the cable's end.

7.8 Costs

On account of their lengthy manufacturing process, braided shields are the most expensive component. A cable with a braided shield may cost twice as much as an unshielded version. Foils shields are significantly less expensive, but not suitable for flexible applications. One should always consider if an inexpensive unshielded cable may be used.

The cost comparison in table 6.2 should provide both a guideline and an incentive for choosing the appropriate cable. For an increasing number of conductors, the proportional cost of the shield diminishes and the factors in the table take on a somewhat different relationship: (1/1.5/2.5/3.2).

7.9 Guarding

If no shield is available, a certain shielding effect can be achieved through guarding. For this purpose, one or two conductors are laid alongside the signal cable requiring protection and connected to either earth or signal ground at both ends. This measure reduces inductive coupling with neighboring signal cables. Within a multi-conductor cable, one of the insulated conductors may, for example, be used to guard a neighboring conductor.

Since the guard conductors tend to act as a shield, they also positively influence the near end crosstalk attenuation. Simultaneously, the inductivity of the cable is reduced, since an opposing electromotive force EMF is built up within the conductors generating compensation current, which in turn induce an opposing magnetic field.

7.10 Shields for Power Supply and Other Insulated Cables

Shields are frequently required for power supply cables as protection against electromagnetic influences. Generally, the same evaluation principles as for signal transmission cables are valid. Although, one additional effect occurring only with power cables should be examined. The high currents create strong magnetic fields which go uncompensated in unbalanced cables. Only a double-sided earthed shield can reduce the magnetic influences. In order to generate an opposing magnetic field that is strong enough, a larger current must be allowed to flow within the shielding circuit. This implies that the entire shielding circuit should have a low resistance/impedance. The shield's cross section should be of sufficient dimension so that the necessary current may flow without unduly heating the shield. This consideration is especially significant for power and three-phase cables having an unbalanced load. Additionally, the current-carrying capacity of the shield must be taken into account for the influences of short-circuit, potential equalization, fault-to-earth and large start-up currents.

8 Standards

National standards and norms exist for cabling techniques as well as for signal and data transmission cables for building installation technologies (see appendix).

8.1 Standards and Specifications for Cables

The German standards for cables are DIN VDE 0245, DIN VDE 0250, DIN VDE 0282, DIN VDE 0815 and DIN VDE 0816. However, these standards define only a few cables for data and signal transmission within building installations. Standards describing the specifications for cables meeting industry standards or system norms are currently under draft.

The international cable standards include: from IEC 189-1 to IEC 189-7, from IEC 708-1 to IEC 708-4 and the draft versions prEN 50167, prEN 50168 and prEN 50169.

Requirement profiles for transmission cables may be found in DIN VDE 0829, DIN 19 245, DIN 19 258, prEN 50098-1, DIN EN 50 173, CLC/ TC 105/WG8-TR 2.0, EIA/TIA-232-E, EIA 485, IEEE 488.2, IEEE 488.1, ISO/IEC DIS 11801 etc.

Further profiles describe diverse industry standards such as IBM cable Types 1, 2, 3 etc.

8.2 Standards for Electromagnetic Compatibility EMC

Standards directly establishing the cable's construction or features do not exist. Whether and how a cable may be disturbed is dependent entirely upon its application. Standards stipulating the permissible interference limits are relevant for cable users. These refer to equipment, systems or other electrical installations, and thereby, indirectly also to cables. The relevant norms are the basis for the EMC regulations valid as of 1.1.1996 without restriction. From this point in time onward, every manufacturer or installer must certify the EMC of his electrical equipment or installations. Non-compliance to a relevant EMC norm is a violation against a valid law for which penalties exist.

8.3 Standards and Guidelines for the Electromagnetic Compatibility of Cables

The following text lists several norms and guidelines which are important for the construction and application of insulated cables.

IEC 1000-x-x   Within TC77 of IEC, the IEC 1000 series of standards were written as a basis for all product standards and activities. They address among others the following topics:
- EMC environments
- limits
- test and measurement procedures
- Installation guidelines and measures

IEC 801-4   Tests according to this standard indicate the highest loading limits for LAN cables for constant, sporadic or periodic interferences.

EN 55011 (DIN VDE 0875 Part 11)   This norm establishes limits and measurement techniques for radio interference arising from industrial, scientific and medical high frequency equipment (ISM equipment).

EN 55022   This standard corresponds to DIN VDE 0878 Part 3: Limits and Measurement techniques for Radio Interference Arising from Information Processing Devices (ITE). It provides measurement techniques to determine the radiated energy of a cable during simulated operation. Furthermore, the disturbance classifications A and B for radio interference voltages are defined.

Within *FTZ TL-6145–3000* of the German Research and Technology Center for the Postal Services, additional information concerning this topic can be found.

9 Selecting Cables
for Signal and Data Transmission
within Building Installations

The cables should be viewed as components of an electronic circuit and not as independent passive components. Fundamentally, the cable's permissible data rates should at least meet the minimum required value. Additionally, the transmission reliability within the anticipated environment should be sufficiently high. The transmission errors may not exceed the tolerable quantity.

Normally, only the unmodulated transmission signal is used for building systems technology. Therefore, coaxial cables are only employed in circumstances where it appears to be unavoidable, because they have complex terminations and provide only one transmission channel for unmodulated signals. Comparatively, a balanced cable can, depending upon the number of pairs, offer many channels. The disadvantages of coaxial cables are not sufficiently compensated by their low wave attenuation. For building installation applications, the usual transmission distances rarely necessitate such high quality attenuation rates. The exception being television antenna cables.

Fiber optic cables are only advantageous for special situations. Larger fiber optic systems for building installations have been until now too expensive. Thus, the application of symmetrical transmission within building system technology is supported. Cables exhibiting balanced characteristics over the entire planned frequency range are necessary. This requirement is also valid for the termination and connection technology employed. One should remember that the weakest link determines the effectiveness of the entire system.

9.1 Symmetrical Transmission Path

During planning, one should strive to avoid differential-mode interference and restrict common-mode interference to a specific tolerance range. The required transmission distances and rates, voltages and signal-to-noise ratio are defined within the specification for the electrical system (see table 1.1 and section 1.2.4). The maximum loop resistance, characteristic impedance and environment class should also be defined in the system specification (section 5.1).

A good balanced cable has :

▷ either the parameters of ISO/IEC DIS 11801 or those established by another norm (section 6.5)

▷ low wave attenuation

▷ high crosstalk attenuation

▷ a large attenuation-to-crosstalk-loss-ratio ACR

▷ a low unbalanced resistance between the conductors of a pair

▷ a low unbalanced capacitance between the conductors and earth

▷ a low mutual capacitance

▷ a high structural return loss

▷ the shortest possible symmetrical length of lay for the stranded elements of a cable to provide high noise immunity against magnetic influences.

The values of these parameters may deviate only slightly along the entire length of the cable. Otherwise, the actual values for shorter cable lengths may strongly differ from the measured or rated values. Rated values usually refer to much longer cable lengths. Other electromechanical cable data should also be defined in the system specification.

The advantage of symmetrical stranding results from the following relationship. A disturbing magnetic alternating field cuts through all enclosed surfaces formed by the stranded conductors. The partial EMF (*E*lectro-*M*otive *F*orce) induced within the individual conductor sections is directly proportional too the length of this section. Within an enclosed surface, the partial EMF induced within the individual sections have opposing directions (diagram 9.1). If the conductor sections forming an enclosed

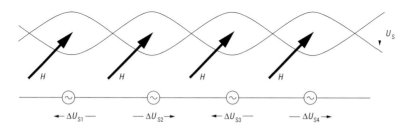

H Alternating magnetic field ΔU_{Sn} Induced voltages
U_S Sum of the voltages

Diagram 9.1 The effect of magnetic fields on symmetrically stranded conductors

surface have identical lengths, then all partial EMFs cancel each other out. No interfering differential voltages ΔU_{Sn} arise at the conductor ends of a section. The sum U_S of all the differential voltages ΔU_{Sn} also equals zero. Thus, no net interference voltage is noticeable at the cable's end.

The stranded sections must also be of equal length and the enclosed surfaces as small as possible, in order to minimize the effects of possible imbalances. This can be achieved through very short stranding lengths (approx. 10 to 30 mm). In order to prevent mechanical deformations of the stranding formation, the stranded assembly should be banded with a braided or foil shield.

An absolutely balanced cable construction which is immune to external magnetic influences, is at least theoretically possible. However, manufacturing and material tolerances lead to a certain dissymmetry. This unbalanced nature can be noticed to varying degrees depending on the frequency range. The signal sources and sinks also do not behave absolutely symmetrical over the entire frequency range. They are especially inclined to greater dissymmetries at high frequencies which result in common-mode to normal-mode conversions (section 4.3).

Within a circuit only conductors of the same pair may be used. If a shield is present, it may only be used for signal transmission. Shields should be appropriately connected to either earth or signal ground (sections 7.6 and 7.7). Interfering ground loops should be avoided.

A single-sided earthed cable shield only protects against low frequency electrical interference fields, high frequency magnetic interference fields and high frequency electromagnetic radiation (sections 7.2.1, 7.2.2 and 7.2.3). Since low frequency magnetic fields usually dominate, it is recommended to better thwart them with a short length of lay (diagram 9.2) than with the added expenditure of shielded cables. As it is too often necessary to sacrifice a double-sided earthing of the shield to avoid ground loops.

The cable trays also influence the EMC of the transmission path. They should be placed far away from interference sources so that only weak

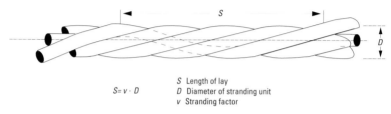

$$S = v \cdot D$$

S Length of lay
D Diameter of stranding unit
v Stranding factor

Diagram 9.2 Length of lay

fields remain present. Their separation from all electrically conductive structures such as steel beams should be a small as possible, in order to minimize common-mode interference voltages due to inductive or capacitive, alternating interference fields. Additionally, the influence of electromagnetic field components which are perpendicular to the cable's axis are, thereby, minimized.

All planning mistakes and deficits are noticeable during or after commissioning and at the latest during final installation, as sporadic interfering common-mode voltages or overly high differential-mode voltages result. With multiple failures, it is very difficult, expensive and often unsuccessful to find the cause within the system, since only two failure symptoms are available. The cable is only one possible source of interference among many.

For the practitioner, it is desirable to be able to derive the cable's rated values from the transmission requirements and the installation environment. Unfortunately such data is not available. The norm ISO/IEC DIS 11801 indirectly contains such an arrangement through its establishment of and correlation between cable categories, classes and cable specifications.

9.2 Unsymmetrical Transmission Paths

The signal sources and sinks of an unsymmetrical transmission path have unsymmetrical characteristics. The cables are unbalanced and the transmission method is consequently also unsymmetrical (section 1.3.1).

The differing lengths of the outgoing and return conductor generate unequal ohmic voltage drops which create differential interference voltages within unbalanced cables. Between the different lengths of the outgoing and returning conductors of a circuit, there is only one large, enclosed (projection) surface which is inevitable due to the cable's construction. Hereby, a magnetic field always generates EMFs of varying magnitude in both conductors which rarely compensate each other. Only complicated, expensive shield constructions can be used to weaken these magnetic fields. In order to reduce the influences of low frequency, alternating magnetic fields, a double-sided shielding is necessary. However, this often leads to galvanic coupling due to the compensation currents flowing through the ground loops. If the shield is simultaneously employed as a common return conductor for several circuits, then additional galvanic coupling due to the interfering currents flowing through the shield occurs alongside the galvanic coupling between the circuits.

The near end crosstalk loss cannot be predetermined, as it is solely dependent upon the actual uses of the individual conductors (layout) for the varying circuits.

As with balanced cables, the EMC of a transmission path can be influenced by the cable's installation. The following points should be considered when defining the use of the individual conductors of a multi-conductor cable:

▷ The outgoing and returning conductors of a circuit should be as physically close to one another as possible within the cable. The enclosed surface for a magnetic field is, thereby, reduced.

▷ A single-ended connection of additional conductors with earth (e. g. as within a balanced cable) improves the balance of the cable.

▷ The shield should only be connected to earth and not to the reference potential.

▷ The EMC can be further improved by shielding the cable with a double shield and by using a ferromagnetic material.

9.3 Coding Signals

The coding method (ciphering) applied to signals influences the noise immunity even for similar transmission rates.

The content representation of the data or signal generally ensues as a temporal sequence of electrical impulses. The correlation between the content and the representation is the code. Coding the same data by various methods leads to differing spectral characteristic curves of the signal voltages. At every point in time, a different frequency spectrum results on the transmission cable depending upon the code, the occupied capacity of the transmission network and the data. The selection of the coding method determines the center of the frequency spectrum. To improve EMC, the spectral center should lie in the lowest possible frequency range, for example MTL-3-Code. Dicing the signal into data packets can reduce the spectral frequency center.

EMC experiments on cables under real operational conditions provided interesting results. The radio interference voltages and bit error rates were measured for various transmission systems and data services (e. g. Ethernet, token ring, CDDI, etc.) at similar frequencies. The data services employ different coding methods. The following paragraphs summarize the experimental results.

Electromagnetic emissions

Within the EU, the radio interference frequencies must be measured for frequency ranges up to 30 MHz. The permissible interference emissions of a system are defined in the pamphlet from the German Federal Ministry for Postal Services and Telecommunications within the provision

243/1991 and in EN 55011. The experimental results showed that the data services possessing their various codes emit very different interference waves. Depending upon the code applied, unshielded cables clearly exceeded the set limits. Shielded and double-ended earthed cables both demonstrated considerably better results. For these cables, the differences between the data services were no longer as drastic. A single-sided earthed cable shield gave less satisfying results than unshielded cables, due to the antenna-like effects of the shield. Cables with non-earthed shields behaved as unshielded cables.

Electromagnetic susceptibility

As anticipated, the unshielded balanced cables were markedly more susceptible to electromagnetic interference than shielded versions. In this case, the shields were earthed at both ends. Cables with non-earthed shields presented somewhat the same results as unshielded cables. Single-sided earthed shields showed yet a worse response than unshielded cables, due to the antenna-like effects of the shield. Otherwise, the results indicate that identical interference influences upon various codes lead to different bit error rates. The susceptibility to electromagnetic interference of Ethernet 10 Base T was especially critical.

9.4 Cables Types, Transmission Modes, Shield Terminations and EMC

The interfering effects of various alternating fields are compared in diagram 9.3.

The surrounding environment disturbs the transmission system with the interference current I, and the electrical interference fields E_1 and E_2 and the magnetic interference field H. E_1 and H stand at right angles to one another, and E_2 is parallel to the cable's axis. The magnitude of the balanced interference voltage U_S induced by its corresponding disturbance field is represented by a continuous arrow, and the unbalanced interference voltage U_{US} of each conductor as a dotted arrow. The length of the arrow is proportional to the value of the interference voltage. As long as the unbalanced interference voltages do not exceed the permissible limits of the signal receiver, only the value of the balanced interference voltage must be considered. The transmission paths depicted in the diagram are only suitable for certain applications.

Diagram 9.3a shows an unsymmetrical transmission with single-conductor, unshielded cables. The return connection is achieved via a common ground. This transmission is only suitable for very insensitive circuits e.g. to control relay coils.

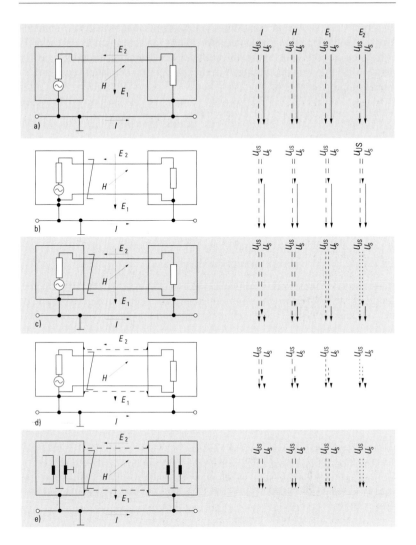

U_S Balanced interference voltage between the conductors
U_{US} Unbalanced interference voltage between a conductor and the reference potential
E_1, E_2 Electrical interference fields
H Magnetic interference field
I Interference current in the reference system (signal ground)

Diagram 9.3   The effects of identical interfering influences in various transmission paths

(Source: Wilhelm, Joh. et al.; Elektromagnetische Verträglichkeit (EMV))

Diagram 9.3 b shows an unsymmetrical transmission with a two-conductor, unshielded cable. The conductor acting as a reference potential is connected at both ends to signal ground. The efforts expended are nearly identical to the transmission path in diagram 9.3 c. The balanced interference voltage is however much larger.

Diagram 9.3 c depicts an unsymmetrical transmission with a two-conductor, unshielded cable. The conductor acting as a reference potential is connected at one end, at the transmitter, to signal ground. This transmission mode is only for analog and digital signal transmission within a weakly contaminated environment.

Diagram 9.3 d depicts an unsymmetrical transmission with a two-conductor, shielded cable. The conductor acting as a reference potential is connected at one end, at the transmitter, to signal ground. Both conductors are surrounded by a common shield which is connected to signal ground at both sides. This transmission mode is suitable for analog or digital signal transmission within a strongly contaminated environment.

Diagram 9.3 e depicts an symmetrical transmission with a two-conductor, shielded cable. The circuit is symmetrically connected to signal ground at the transmitter. Both conductors are surrounded by a common shield which is connected to signal ground at both sides. This transmission mode is suitable for analog or digital signal transmission within a strongly contaminated environment.

As the comparison in diagrams 9.3 c, 9.3 d and 9.3 e shows, an additional cable shield for unsymmetrical transmission is only helpful in reducing the less critical, unbalanced interference voltages. The balanced interference voltage is still considerably larger than for a symmetrical transmission with a shielded cable.

For this comparison, the peculiarities of unsymmetrical transmission for multi-conductor cables, such as a lower near end crosstalk attenuation and galvanic coupling via common return conductors, were not evaluated.

9.5 Fiber Optic Technology within Building Installations

The conventional networks for data and signal transmission in buildings are common distribution networks. This implies shorter distances, higher transmission rates and a very large number of participants. Optical transmission has two disadvantages for line- or tree-shaped cabling structures. The light pulse does not propagate in a passive node with equal intensity in all directions, but rather is weakened depending on the direction and the number of branches. The light pulse no longer meets the requirements for building installation technology. Active distributions are presently still too expensive, especially when considering the large numbers involved.

Additionally, mineral glass fiber optic cables and their complex installation and connection techniques are too expensive. In the future, cost effective star-couplers for a star-shaped bus structure will be available.

For the general application of fiber optic cables in building installations, all data services offered within the building must also be suitable for fiber optics. Otherwise, additional copper cables must still be laid in parallel to the fiber optic cables, since all data services should be accessible from every port.

Coarsely linked networks based on fiber optics are already employed in industrial areas. Large quantities of data must often be transmitted between office buildings and workshops or within a building between offices and machinery (e.g. CAD data, data for numerical programs, etc.). It is often favorable with regard to EMC, to forego copper cables and choose a more expensive fiber optic transmission system. Conventional polymer optical fibers have a limited transmission length, therefore, they can not be employed in a distribution network.

In comparison to copper networks, additional expenditures for the electro-optic converters as well as for eventual active fiber optic distributors should be considered. Furthermore, fiber optic networks require additional copper cables to supply power to the communication modules. In such instances, relatively little or no savings from a reduction in copper cables result.

Presently, there is no network norm recommending polymer optical fibers. For the present technological stage of development, it is only sensible to apply fiber optics within building installations, where extremely high EMC demands are placed on the transmission system (i.e. secure information transmission without interception).

For point-to-point transmission, the situation is naturally viewed from an entirely different angle. Fiber optic technology is especially advantageous for these paths. For example:

▷ building links

▷ links between computers and/or their peripheral equipment

▷ traversing through rooms where the installation of copper cables is prohibited

▷ transmission paths placing weight and volume restrictions on the cables

▷ the extension of a transmission path which cannot be realized due to the transmission characteristics by copper cables (e.g. CENTRONICS interface over long distances).

For typical transmission distances in buildings, the following transmission rates are presently used:

Polymer optical fibers	10 MHz
Polymer-cladded glass fibers	70 MHz
Mineral glass fibers (graded index profile)	>100 MHz

An important parameter for fiber optics is the bandwidth-length product. It provides a measure of the dispersion within the fiber. The larger the value of the product, the higher the transmission frequencies, and the greater the bridgeable transmission distances. The presently valid magnitudes are listed in table 9.1.

Due to the physical limitations of the materials presently employed in polymer optical fibers, no appreciable improvements in the attenuation constants are foreseeable in the near future.

Since fiber optic cables must be terminated on-site at the building location, the installation techniques must be fast, simple and reliable. Unfortunately, insufficient standards exist for the connectors and sockets of polymer optical fibers. On the other hand, the installation and connection technology for polymer optical fibers is simple, inexpensive and highly reproducible. On building sites, they offer enormous advantages in comparison to mineral glass fiber optics.

The many different diameters for optical fibers are critical. Every diameter and every cable type may eventually require its own special version of connectors and sockets as well as links to the communication subscribers. It is naturally advantageous on the building site to reach for a modular fiber optic connection system. Whereby, one fiber optic connector may be employed for fibers of any diameter. The most cost-effective, fiber optic cable could then be selected for the transmission distance at hand. For example, complete V.24/RS232 interface modules with electro-optical converters are available to which a variety of cables (i.e. polymer optical fiber, mineral glass, polymer-cladded glass fibers, etc.) may be connected.

Table 9.1
The product bandwidth (MHz) times length (km) of various optical fibers

Polymer optical fibers	typical	10 MHz 1 km	at 660 nm
Polymer-cladded glass fibers	min.	17 MHz 1 km	at 820 nm
Mineral glass multimode fibers, 50 μm,	min.	200 MHz 1 km	at 820 nm
Graded index profile	min.	600 MHz 1 km	at 1300 nm
Mineral glass multimode fibers, 62.5 μm,	min.	400 MHz 1 km	at 820 nm
Graded index profile	min.	600 MHz 1 km	at 1300 nm

The dimensions given in nm represent the wave length of the light used for transmission.

In general, the following times are necessary for connector installation:

Mineral glass fiber optics   approx. 20 . . . 30 min
(Graded index profile)

Polymer cladded fiber optics   approx. 5 min ("Crimp & Cleave"
                                                   Technology)

Polymer optical fibers   approx. 2 min ("Hot Plate" Technology).

Polishing the face of the fiber is only necessary for mineral glass fibers.

Table 9.2 contains dimensions for the V.24/RS232 fiber optic system.

Table 9.2   Standard values for V.24/RS232 fiber optic systems

	Max. transmission distance approx. m	Core diameter of optical fiber μm	Max. transmission rate* kbit/s
Polymer optical fiber system	40 or 50**	980	max. 38.4
Polymer-cladded fiber optic system	1100 . . . 2100	200	max. 38.4
Mineral glass fiber optic system	2000 . . . 15000	62.5 (50)	max. 38.4

  * The maximum transmission rate is determined by the electronic module and not the
    optical fiber.
  ** The maximum transmission distance of the polymer optical fiber is determined by the
     power limitations of the available optical transmitters (LEDs).

Power supply for the electronics may alternatively be provided by an external power supply over an additional connector or over a free pin offered by the interface connector. Polymer optical fiber systems offer the possibility of supplying power directly from the data signal, making additional power supplies superfluous.

Optical fibers

The core of a polymer optical fiber consists of a single fiber core made of polymethylmethacrylate (PMMA). It is coated with a fluoridated PMMA layer. The core diameter for the popular step index profile fiber is 980 μm. An additional covering is added for mechanical protection. The outer diameter is 1 mm (diagram 9.4). The optical window of the PMMA fiber lies at approx. 650 nm. An attenuation value of less than 200 dB/km is achieved. Even attenuation values of 160 dB/km are possible. Tensile strain and pressure increase the attenuation of the fibers, just as with other fiber materials. As long as the destructive limits are not crossed, these effects are reversible.

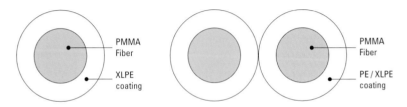

Diagram 9.4   Cables with polymer optical fibers

Diagram 9.4 shows a cable with polymer optical fibers which is suitable for fixed installation. The manufacturer provides the following data:

Temperature range	−55 °C ... +70 °C (Type A01)
	−55 °C ... +115 °C (Version A0X)
Permissible tensile strain	max. 5 N (continuous)
	max. 35 N (intermittent)
Resistance to transverse pressure	max. 30 N/cm (continuous)
	max. 200 N/cm (intermittent)
Bending radius	min. 30 mm (continuous)
	min. 10 mm (intermittent)
Attenuation	max. 200 dB/km (Version A01)
	max. 400 dB/km (Version A0X)
Refractive index of core	1.492
Refractive index of the PMMA layer	1.417
Numerical aperture	0.47
Angle of acceptance	56°

Polymer-cladded glass fibers (PCF)

Polymer-cladded glass fibers (PCF) (diagram 9.5) have a 200 μm thick mineral glass core coated with a polymer layer (cladding). The light waves are reflected at the cladding. This construction gives a step-index profile. The outer diameter over the cladding is 230 μm. A protective covering called a buffer is brought over the cladding. The optical window lies approx. between 600 and 850 nm and intersects with the window for polymer optical fibers. The attenuation at 660 nm is approx. 7 dB/km. In spite of a high insertion loss (coupling attenuation), transmission distances of up to 500 m have been realized. PCF fibers are less expensive than mineral glass fibers yet more expensive than polymer optical fibers.

For a duplex cable using polymer-cladded glass fibers, the following technical data is generally given:

Temperature range	−40 °C ... 85 °C
Permissible tensile strain	max. 90 N (continuous)
	max. 270 N (intermittent)

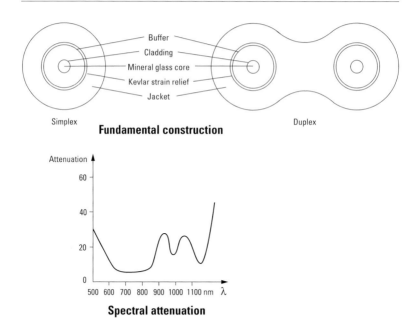

Fundamental construction

Simplex | Duplex

Buffer
Cladding
Mineral glass core
Kevlar strain relief
Jacket

Spectral attenuation

Diagram 9.5   Cables with polymer-cladded glass fibers

Bending radius	min. 25 mm (during installation)
	min. 15 mm (during operating)
Attenuation	max. 7 dB/km (at a wavelength of 660 nm)
	max. 6 dB/km (at a wavelength of 820 nm)
Numerical aperture	0.36
Core diameter	200 μm
Diameter over cladding	230 μm

Mineral glass fiber optics

Mineral glass fiber optics are the original fiber optic transmission medium. Its developmental stage is by far the most advanced and the variety of models is extensive. A description of them is beyond the scope of this book. It suffices here to mention that mineral glass fiber technology was primarily developed for long-distance telecommunication networks. The technical efforts are characterized by this nature. The network standards for cabling within building installations mostly allow for optical fibers as building links or for floor distribution. Their general intended application in FDDI (*F*iber *D*ata *D*istributed *I*nterface) will gain acceptance at a future date. Balanced copper cables can, in the meantime, reliably transmit up to 600 MHz over the typical distances found within

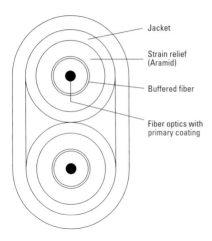

Diagram 9.6 Indoor cable with mineral glass fiber optics

buildings. Diagram 9.6 shows an indoor cable with two mineral glass optical fibers.

The most important technical data are:

Fiber diameter   62.5 μm
Profile   Graded index
Temperature range: −10 °C ... +50 °C (during installation)
                             −20 °C ... +50 °C (during operation)
Permissible tensile strain   max. 360 N (continuous)
                             max. 600 N (intermittent)
Bending radius   min. 50 mm (during installation)
                             min. 80 mm (during operation)
Resistance to transverse pressure   max. 50 N/cm (continuous)
Resistance to alternate bending   min. 6000 Cycles
Attenuation   max. 3.5 dB/km
                             (at a wavelength of 850 nm)
                             max. 0.8 dB/km
                             (at a wavelength of 1300 nm)
Numerical aperture   0.36
Core diameter   200 μm
Jacket diameter   230 μm

10 Installation of Cables

The installation of high quality data and signal transmission cables is no longer comparable to the installation of power supply cables. Requirements which maintain the cable's relevant transmission characteristics and EMC must be observed. Both the cable's attributes and EMC react extremely sensitively to mechanical and thermal influences.

At transmission frequencies of 100 MHz or more, installation errors are clearly evident. They can deteriorate the characteristic of the entire transmission path to such an extent, that the planned transmission rates can hardly be achieved. Unfortunately, this poor performance is often first observed when the network is fully extended or operating at full capacity. If the maximum number of participants or the maximum length of cabling is used, eventual network disturbances are especially painful. Improvements to the network are very costly and difficult. Often only the installation of a new network is helpful.

Only high quality, balanced cables are suitable for high frequency symmetrical transmission. Their EMC and transmission characteristics are largely determined by their balanced attributes. These should not be degraded during installation.

The balance of a cable is good, when the geometrical arrangement of the cable's components within the cross section and along the cable is absolutely symmetrical and meets the rated values within a tight tolerance range. Pressure, tensile strain or bending can alter the cable's balance. One should discern between the following cases:

▷ reversible changes (elastic range of the materials)

▷ irreversible changes (lasting deformations remain due to changes in the structure of the materials) and

▷ destructive changes (breaks, fissures, cracks).

Shields or other wrappings usually lend the cable a higher threshold to mechanical stresses.

10.1 Tensile Strain

Tensile strain occurs along the longitudinal axis of the cable. When the maximum permissible value stated by the manufacturer is exceeded, the conductor resistance increases and consequently the attenuation. If the attenuation in the various conductors increases non-uniformly, the unbalanced resistance also rises. Furthermore, the cable's diameter decreases. Consequently, the cable's construction becomes more compact leading to higher capacities and attenuation constants. Due to these alterations the characteristic impedance also sinks.

The permissible tensile strain refers to the summation of all conductor cross sectional areas. The other cable components do not carry any load. According to DIN VDE 0298 Part 3 the maximum permissible tensile stress F may not be greater than

$$F = (\text{Sum of all conductor cross sections in mm}^2) \cdot 50 \text{ N/mm}^2. \qquad (8)$$

10.2 Bending Stress

Bending stress (flexural load) leads to tensile stress on the outer radius of the bent cable and to transverse pressure on the inner radius of the bent cable. Both shift the geometrical position of the cable's components in relation to one another. Consequently, the near end crosstalk loss (*NEXT*) deteriorates. The permissible minimum bending radius may, therefore, not be exceeded at any time including during installation.

10.3 Torsional Load

Twisting the cable along its longitudinal axis results in alterations for the length of lay of the stranded conductors. The conductors are also shifted in relation to one another. The near-end crosstalk ratio deteriorates, and the characteristic impedance along the cable is changed. Disturbing reflections during operation result, as the structural return loss α_r of the cable is reduced. Fundamentally, torsion must be avoided. Improper unreeling of the cable is the prime culprit. It is a huge installation mistake to pull the cable up over the top of the spool in order to unwind it.

10.4 Strain due to Pressure

Transverse pressure occurs perpendicular to the cable's axis on its surface. Hereby, the geometries of the components are deformed as well as their position with respect to each other or the cable's axis. This stress changes the characteristic impedance. If the pressure occurs at discrete locations, reflections result at these points. A major culprit can be the clamps used to fasten a cable at regular intervals along the cable. In the extreme case, it may result in total reflection of the signal.

10.5 Temperature Effects

The thermoplastic materials of the cable's components begin to soften at increased temperatures. This reduces their robustness to mechanical stress. It can lead to lasting deformations of the components. In the extreme case, the insulation may be penetrated or crushed resulting in short circuit conditions. On the other hand, low temperatures decrease the elasticity of the materials. Therefor, the outer jacket may crack during bending. For these reasons, the temperature limits may not be exceeded.

Diagram 10.1 shows that the dominant compounds used for telecommunications cables (PE for insulation and jackets, and PVC for jackets) react sensitively to mechanical pressure at even small excursions outside of the permissible temperature range. At a compound temperature of 110 °C, the test sample of LDPE-insulation flowed completely out from under the test stamp after only four hours.

The test for the thermal-pressure relationship must be performed on an insulated, single-conductor cable according to DIN VDE 0472.

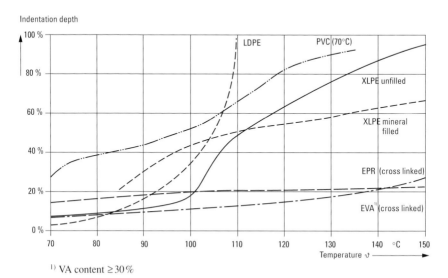

Indentation depth

1) VA content ≥ 30%

Diagram 10.1 Thermal-pressure relationship of polyolefines

(Source: Heinhold, Lothar; Kabel and Leitungen für Starkstrom – Teil 1)

10.6 Mechanical Compatibility

As previously mentioned, the uniformity of the transmission characteristics and EMC along the cable should be strived towards. The construction of the clamps chosen to fasten the cables during installation must not hinder this uniformity. Only then may the conductors and shields be properly terminated.

The type of fastening technology chosen along with the mechanical construction of the cables directly influences the necessary installation efforts. Unfavorable combinations may result in very long installation times or indeterminable quality.

11 The Future of Building Systems Technology

In the future, local networks will be increasingly included in building installation technologies. The growing demands of the building's occupant and the desire of the owner to ensure the building's capability to meet future demands are reasons for the installation of local area networks. Therefore, data services, control and security systems must continually satisfy the growing demands for speed, efficiency, reliability when transmitting data and signals. Future industrial control technology will also operate at much higher frequencies than those presently used.

It remains important that the cabling can be modified and expanded. The high follow-up costs for alterations in the cabling system should be reviewed. The signal and data cables themselves only comprise approx. 3 to 5 % of the entire expense for the complete cabling system within a building. Therefore, a non-recurring, initial cost savings here should not be chosen at the expense of a future-oriented, reliable system featuring high quality transmission performance. Careful planning, the use of high quality balanced cables, compatible components and a qualified installation method are the only guarantee for low follow-up (hidden) costs and a flexible, future-oriented cabling system.

Fiber optics should be employed: where high EMC requirements exist, where immunity to lightening disturbances are necessary, for applications in high voltage plants, for control applications which cannot be satisfied with copper cables.

When considering all of the fundamental transmission principles and the facets of EMC, copper cables remain the best choice for industrial control applications, drive technology and building systems technology.

12 Appendices

12.1 Standards

The German standards and norms do not comprehensively and sufficiently support the planners and designers of signal transmission equipment and systems. Therefore, the international norms are of equivalent importance.

The following defines common abbreviations:

ANSI	American National Standards Institute
DIN	Deutsche Industrie Norm
E	Draft (Entwurf)
EIA	ELECTRONIC INDUSTRIES ASSOCIATION
EN	European Norm
HD	Harmonised Document
IEC	International Electrotechnical Commission
IEEE	Institute of Electrical and Electronics Engineers
ISO	International Organization for Standardization
ITU	International Telecommunication Union
NEMA	National Electrical Manufacturers Association
pr	Draft of European Norm
Std	Standard
T	Part
TIA	TELECOMMUNICATIONS INDUSTRY ASSOCIATION
SB	Technical Systems Bulletin
V	Preliminary standard
VDE	Verband Deutscher Elektrotechniker
ZVEI	Zentralverband Elektrotechnik- and Elektroindustrie e.V.
ZVEH	Zentralverband der Deutschen Elektrohandwerke

12.1.1 National Standards

DIN 19245	PROFIBUS, Process Field Bus
DIN 19258	INTERBUS-S, Sensor-/Aktorennetzwerk für industrielle Steuerungssysteme
DIN EN 50081	Elektromagnetische Verträglichkeit (EMV); Fachgrundnorm Störaussendungen/Störfestigkeit
E DIN EN 50210	Schutz von Telekommunikationsleitungen gegen atmosphärische Entladungen Lichtwellenleiteranlagen

E DIN EN 50217	Störaussendungsmessungen am Aufstellungsort
DIN 47250	Hochfrequenz (HF)-Kabel und -Leitungen
DIN VDE 0100	Errichten von Starkstromanlagen mit Nennspannungen bis 1000 V
E DIN VDE 0245	Leitungen für elektrische und elektronische Betriebsmittel in Starkstromanlagen
DIN VDE 0472	Prüfung an Kabeln and isolierten Leitungen
DIN VDE 0800	Fernmeldetechnik
DIN VDE 0815	Installationskabel und -leitungen für Fernmelde- and Informationsverarbeitungsanlagen
DIN VDE 0819	Fachgrundspezifikation für mehradrige und symmetrische paar-/viererverseilte Kabel für digitale Nachrichtenübertragung
DIN V VDE 0829	Elektrische Systemtechnik für Heim und Gebäude (EHSG)
DIN VDE 0843	Elektromagnetische Verträglichkeit von Meß-, Steuer- and Regeleinrichtungen in der industriellen Prozeßtechnik
DIN VDE 0871 T 11	Funkstörgrenzwerte and Meßverfahren
E DIN VDE 0877-1	Messen von Funkstörungen Messen von Funkstörspannungen
DIN VDE 0878	Funkentstörung von Anlagen and Geräten der Fernmeldetechnik
DIN VDE 0891	Verwendung von Kabeln and isolierten Leitungen für Fernmeldeanlagen and Informationsverarbeitungsanlagen

12.1.2 International Standards

ANSI/IEEE Std 488.1	IEEE Standard Digital Interface for Programmable Instrumentation
EIA/TIA TSB-36	Technical Systems Bulletin Additional Cable Specifications for Unshielded Twisted-Pair Cables
EIA/TIA TSB-40 A	TIA/EIA Telecommunications Systems Bulletin Additional Transmission Specifications for Unshielded Twisted-Pair Connecting Hardware

EIA/TIA-232-E	Interface Between Data Terminal Equipment and Data Circuit-Terminating Equipment Employing Serial Binary Data Interchange
EIA/TIA 485	Standard for Electrical Characteristics of Generators and Receivers for Use in Balanced Digital Multipoint Systems
EIA/TIA-568	Commercial Building Telecommunications Wiring Standard
EN 50090	Elektrische Systemtechnik für Heim and Gebäude (ESHG)
E prEN 50097	Medizinische elektrische Geräte: Elektromagnetische Verträglichkeit
EN 50098	Informationstechnische Verkabelung von Gebäudekomplexen
EN 50167	Bauartspezifikation für Etagenkabel mit gemeinsamen Schirm für digitale Kommunikation
EN 50168	Bauartspezifikation für Geräteanschlußkabel mit gemeinsamen Schirm für digitale Kommunikation
EN 50169	Bauartspezifikation für Verteilerkabel (Gebäudeverbindungskabel und Steigekabel) mit gemeinsamen Schirm für digitale Kommunikation
EN 50173	Information technology Generic cabling for customer premises cabling
EN 55011	Funkstörgrenzwerte und Meßverfahren
EN 55022	Grenzwerte und Meßverfahren für Funkstörungen von Einrichtungen der Informationstechnik
IEC 189-1/2	Niederfrequenzkabel, Leitungen und Drähte mit PVC-Isolierung und PVC-Mantel
IEC 708-1	Niederfrequenzkabel mit Isolierung und Feuchtigkeitsschutzmantel aus Polyolefinen
IEC 794-1	Lichtwellenleiter-Kabel
IEC 801	Elektromagnetische Verträglichkeit für Meß-, Steuer- und Regeleinrichtungen
IEC 1156-3	Vieladrige, Paar- und Viererverseilte Datenkabel, Part 3: Geräteanschlußkabel
IEC 1156–4	Vieladrige, Paar- und Viererverseilte Datenkabel, Part 4: Steigekabel

IEC 1196	Hochfrequenzkabel
IEEE 488	Digitale Schnittstelle für die programmierbare Instrumentierung
IEEE 1118	Serieller Steuerbus für eine Mikroregeleinrichtung
ISO/IEC DIS 11801	Information technology Generic cabling for customer premises cabling
NEMA	Guide for classification of all types of insulated wire and cable

12.2 Cable Type Designation Codes for Communication and Information Processing Equipment According to DIN VDE (Excerpt)

The type designations for communication cables consist of an alpha-numeric code. The first position indicates either the cable type or its use. The alpha-numeric code for the essential elements of the cable's construction is preceded by a hyphen. The succeeding numerals indicate the number of conductors or pairs times the conductor's size. The cross-sectional area for stranded conductors is given in mm^2; whereas, the diameter of a solid conductor is expressed in mm. The type of stranding is given at the end of the type designation code.

Occasionally, the designations have further meanings. The correct definition is established and defined by the cable's corresponding standard.

Letters for the first position of the type designation code

A-   outdoor cable

J-   installation cable

JE-   installation cable for industrial electronics

L-   flexible cords for communication equipment, cables having finely stranded conductors for higher mechanical stresses within communication equipment

S-   switchboard cables for communication equipment

Type designation code for the elements of the cable's construction

The order of the alpha-numeric code corresponds to the successive layers of the construction elements beginning from the center outwards.

B   armouring

C   braided copper wire shield

109

(C)	shield or outer conductor made of braided copper wire over a pair
E	earthing layer with an embedded layer of plastic tape
F(L)2Y	cable core filled with petrolate having a laminated outer jacket
H	insulation or outer jacket made from a halogenfree compound
(L)Y	laminated outer jacket made of aluminum tape and PVC
(L)2Y	laminated outer jacket made of aluminum tape and PE
(ST)	static shield of metal tape or plastic-laminated metal foil tape
Y	insulation, jacket or protective covering of PVC
2Y	insulation, jacket or protective covering of PE
02Y	insulation of cellular PE
02YS	insulation of cellular PE with outer layer of non-cellular compound (foam skin)
X	insulation, jacket or protective covering of cross-linked PVC (XLPVC)
2X	insulation, jacket or protective covering of cross-linked PE (XLPE)
(Z)	tensile strain-proof braiding of steel wires

Type designation code of the stranding elements

Bd	unit-type stranding
Bd...	unit-type stranding; ... number of stranding elements in a in unit, bunch or in a designated group
...IMF	individual stranding elements (e.g. pairs = PIMF, cores = AIMF or quad = VIMF) within a metal foil or metallic paper with a drain wire
Lg	stranded in layers
St	star quad with phantom circuit
St I	star quad for long distance cables
St III	star quad for local communication cables

12.3 Frequency Ranges for Mobile Communication Services

Services	Frequency MHz
Bunched radio	410...415 and 420...425
Cityruf; Inforuf; Euromessage	465.97; 466.075; 466.23
ERMES	169.4...169.8
Eurosignal	87.365...87.340
FPLMTS	1900...(2020) and (2110)...2200
Inmarsat	(1510...1530)
Mobil communications	
B-Net (to 31.12.94)	148.21...149.13 and 153.01...153.73
C-Net	450...455.75 and 460...465.97
D-Net	890...915 and 935...960
E1-Net (cordless Tel.) DCS 1800	(1710)...1790 and (1815)...(1880).
MobSat	2500...(2515) and (2640)...(2670)
Modacom	417.01...417.37 and 426.6...427.01
DECT (cordless phone)	(1880)...1900
Telepoint C11 + (birdie)	885...887 and 930...932
Telepoint C12	864 ...868
TFTS	(1670)...(1685) 1800...(1815)

( ) approx. Value

12.4 Checklist for LAN Cables

Presently, there are still various standards and specifications for LAN cables. None of these completely and unambiguously describe a LAN cable, rather, they describe only specific characteristics or test procedures. In order to select an appropriate LAN cable, further application-specific details are necessary:

(Check box □ if applicable.)

1   Standards, Specifications
     (at least one is necessary)

1.1   □[1] ISO/IEC DIS 11801   □ Cable Category 3

         (shielding: optional)   □ Cable Category 4

                                    □ Cable Category 5

                                    □ Cable Category 5, Table 18

1.2   □   EIA/TIA TSB-36   □ Category 3

         (only UTP cable)   □ Category 4

                                    □ Category 5

1.3   □   UL   □ Level I

                                    □ Level II

                                    □ Level III

                                    □ Level IV

                                    □ Level V

2   Characteristic impedance

2.1   □[1] 100 Ω

2.2   □   120 Ω

2.3   □   150 Ω

[1] preferred values

3   Stranding element

3.1   □[1] Pair

3.2   □   Quad

3.3   □   Shielded stranding element

4   Conductor diameter

4.1   □[1] 0.6 mm

4.2   □   ... mm

4.3   □   AWG.../...

UTP: Unshielded Twisted Pair
STP: Shielded Twisted Pair
-S or F: Overall shield

Example

UTP-S or FTP: unshielded, twisted pair, Overall shield

5   Number of stranding elements

5.1   □   1

5.2   □   2

5.3   □   3

5.4   □   ...

6   Overall shield

6.1   □   Foil

6.2   □   Braiding

[1] preferred values

7   Burning behaviour

7.1 □ DIN VDE 0472 Part 804 Test Method A

7.2 □ DIN VDE 0472 Part 804 Test Method B

7.3 □ DIN VDE 0472 Part 804 Test Method C

7.4 □ IEC 332–1(1979) ed 2

7.5 □ IEC 332–2

7.6 □ IEC 332–3

8   Halogen content

8.1 □ DIN VDE 0472 Part 813

8.2 □ ....

9   Functionality in the event of fire

9.1 □ DIN 4102   E 30

                             E 60

                             E 90

                             E 120

9.2 □ ....

10   Special requirements

13 Bibliography

Kohling, A. CE Conformity Marking, 2nd edn. Publicis MCD Verlag, Erlangen, 1996.

Schubert, W. *Nachrichtenkabel and Übertragungssysteme,* 2nd edn. Siemens, Berlin, München, 1980.

Mahlke, G., Gössing, P. *Fiber Optic Cables,* 3rd edn. Publicis MCD, Erlangen, 1997.

Heinhold, L. *Power Cables and their Application,* 3rd edn. Siemens, Berlin, München, 1990.

Seip, G. *Elektrische Installationstechnik,* 3rd edn. Siemens, Berlin, München, 1993.

Schwab, A. J. *Elektromagnetische Verträglichkeit,* 3rd edn. Springer, Berlin, 1993.

Beuth, K. *Digitaltechnik,* 9th edn. Vogel, Würzburg, 1992.

Wilhelm, J. and 9 co-authors. *Elektromagnetische Verträglichkeit (EMV),* 5th edn. expert, Ehingen bei Böblingen, 1992.

Retzlaff, E. *Lexikon der Kurzzeichen für Kabel and isolierte Leitungen,* 4th edn. vde, Berlin Offenbach, 1993.

ZVEI, ZVEH. *Handbuch Gebäudesystemtechnik, EIB,* 2nd edn. Wirtschaftsförderungsgesellschaft der Elektrohandwerke, Frankfurt, 1994.

Seminar documents. *Symmetrische Kabel für zukunftssichere Datennetze.*

Technische Akademie Esslingen Weiterbildungszentrum, Ostfildern, 1994.

Seminar documents. *Elektromagnetische Verträglichkeit (EMV).*

OTTI Ostbayerisches Technologie Transfer Institut, Regensburg, 1994.

Halling, H. *Serielle Busse,* vde.

Forst, H.-J. *Gebäudeautomatisierung.* vde, Berlin, Offenbach.

Pattay v. W. *Informationstechnische Verkabelung von Gebäudekomplexen,* expert, Ehingen bei Böblingen, 1994.

14 Index

Bezner, Heinrich
Dictionary of Power Engineering and Automation
Part 1 German/English
3rd revised and enlarged edition, 1993, 579 pages, 14.8 cm x 22 cm, hardcover
ISBN 3-8009-4118-X
DEM 98.00/öS 715.00/sFr 96.00

Part 2 English/German
3rd revised and enlarged edition, 1993, 511 pages, 14.8 cm x 22 cm, hardcover
ISBN 3-8009-4119-8
DEM 98.00/öS 715.00/sFr 96.00

CD-ROM Version
Edition 1996
System requirements
– Microsoft® Windows™ 3.x, Windows 95
– Personal computer with at least a 386 microprocessor and
– min. 4 MB RAM
– CD-ROM drive
– VGA or higher resolution video adapter
– Mouse or adequate pointing device

ISBN 3-89578-045-6
Price for single user
DEM 229.00/öS 1672.00/sFr 203.00
Price for multi-user on request.

This publication covers terms essentially from the following fields: Power generation, transmission and distribution; drives, switchgear and installation technology, power electronics; measuring and analysis technology, test engineering; automation technology and process control.
With over 65,000 entries in the German-English part of the dictionary and 52,000 in the English-German part, as well as with its many cross-references to VDE codes and standards (DIN, VDI, IEC, BS, ANSI, CEE, ISO), this dictionary is a reliable reference source for all those who read, prepare or process power engineering and automation texts in either language.